HUMANITY IN SPACE

A pre-history of humanity's second century in space

Tim Chrisman

CONTENTS

Introduction

INTRODUCTION

It has been a year since I started this book, so I guess now is as good a time as ever to write the introduction. Since I am not the guy who reads introductions, I can only assume what their purpose is (to explain why I wrote the book that you are now reading). With that in mind to help calibrate your expectations (and my writing) let's go.

Humanity in space. There are few people alive who spent much time without humans being in space. But we don't know what that means. No one tells their kids...oh well I guess if that doesn't work out you can always be an astronaut. Telling a child about going to space is talked about in the same way being President or a fairy princess is. That's because humans in space aren't real...not in our minds.

We know what is real because we can touch it, or because we can envision it. Since humans sat around fires telling legends we have brought the untouchable to life through stories. There were no cavemen explaining the math behind throwing a spear, nor were there any ancient Egyptians talking about the mechanics of biology. We brought those concepts to life by telling a story. Stories about people we all knew doing things we all knew about. Stories anchored the abstract into reality.

This has been the case forever because math, science, and engineering are scary, complicated or just outside what we know. Space is the perfect storm of scary complicated topics. This book is meant to change that. It isn't a science book, nor is it a work of fiction. It is a space-age cave drawing.

There are going to be things that aren't 100% technically accurate in the stories. There are going to be complicated concepts in the explanations of them. But at the end of the day, this book is meant to be a cartoonish rendition of the future. Not cartoonish as in funny, but cartoonish as in an attempt to portray the broad outlines of something without trying to get it perfectly accurate.

If you are still with me then you should know that this book isn't really meant to be read sequentially. I mean it can be, and it makes sense to be. But the book arose out of a blog so each section is independent and able to be read and understood on its own. So, feel free to look at the table of contents and just read the things you are interested in…or don't and look at the back of the book to see if the hero lives or dies… spoiler…there is no hero…other than me for writing this book.

Ok enough intro…hopefully you enjoy reading this as much as I enjoyed writing it.

-Tim

HOW WILL WE
GET THERE?

The Longest Ride Ever

"It was beautiful, there was no other way to look at it," Andy thought, as she gazed at the thousands of glittering objects she could see from her window of the tram climbing its way up the cable to orbit. Intellectually, she knew that most were just small satellites, ranging from broadband internet providers to clusters of what she could only suspect were imaging constellations, which used a machine learning algorithm to stitch the images of each individual sensor into an ultra-high definition image of the surface.[1]

Knowing these things didn't make seeing them up close any less impressive. Andy chuckled to herself, "good thing I am so easily impressed, otherwise this three-day trip would be unbearable." Already she had seen a thunderstorm from the top and a launch from the nearby space launch facility. "I'll bet that launch was one of those classified government programs that was forbidden by law from using the space elevator," she mused.

Andy was headed to orbit onboard the new-ish space elevator as a senior mineral consultant for Asteroid's Extraction (AE), a multi-national mining corporation. AE was one of the big three companies dominating activity outside cislunar space. The low cost of 'playing' in space meant that there were literally thousands of companies vying for attention between the Earth and the Moon, supplying everything from communication, to fuel, to what passed for purified water. However, that very diversity of industry ensured the cost and risk of moving beyond view of Earth were both prohibitive. Only a handful of governments had even launched small teams beyond Mars, and all of those were temporary trips.

As Andy neared orbit, she began to feel what humans have always felt as they gazed on the stars...the limitless possibilities associated with viewing the infinite. 'Maybe one day' she thought, 'maybe one day I'll make it even further out there.'

The Real Deal

A three-day elevator trip, really?

It sounds excruciatingly slow, especially to all of you who already hate getting into a couple minute elevator ride to your hotel room. Remember though – it takes you a couple minutes to get to the top of your hundred, or couple hundred-foot tower. Getting to geosynchronous orbit is going to take just a tad longer.

The Elevator

First off, you are probably wondering (a) what is a space elevator and how do they work.[2]

For starters, they work exactly like the elevator in your hotel does. There is an anchor at the top, connected to one or multiple cars. Those cars are raised or lowered on demand along cabling. It seems simple enough. The main difference between an orbital elevator and a hotel one, is the length of the cable, and the size / placement of the anchor.

A huge problem with the anchor, is that it will not be fixed. After all, there isn't anything in space to 'hold onto'. So, the anchor will have to be an orbital installation of some kind.[3] Its exact size and mass requirements will vary, depending on the amount of cargo it will haul. What won't vary, is the orbit it sits in. An object in low earth orbit (LEO) circumnavigates the globe nearly 20 times a day; while a similar object in mid-earth orbit (MEO) will do so between 2-12 times a day.[4] This is a problem if you want your elevator to start and end at the same place, since, at best, your orbital 'top floor' will only be over the 'ground floor' for an hour or so.

Solving this means putting the elevator anchor in geosynchronous orbit (GEO), which is about 22,236 miles up.[5] Before we turn to the math involved to figure out how long a trip this would be, I should note that it is what's known as a 'long way up'.

The Math

OK – I know some of you are thinking. How will a guy who was on a competitive math team growing up, explain math to us simply? This is a fair concern, since some of my former teammates and coaches went on to be theoretical mathematicians. I, on the other hand, did not. I learned in that experience that (1) math is easier when you have shortcuts, and (2) I wanted the shortest possible path through the math.

Because of those lessons, let's take the short path through the math involved here.

Step 1: Elevator Speed

Q1: How fast does a 'regular' elevator go?

A1: Don't know...let me check google A1a: 15-22 meters per second[6]

Q2: WTF does that mean in American? A2: Uhhh... fast?

A2a (after a google conversion): 50-70 feet per second

Step 2: Height of Top Floor

Q: How high is the top floor

A: Ha! I have that answer, because GEO is at 22,236 miles.

Q2: How many feet is that?

A2: Damn it

A2a (post-calculator): 117,406,080 feet

Step 3: Divide Height by Speed To Get Time

Q: What is 117,406,080 feet divided by 50-70 feet per second?

A: 1,677,229 – 2,348,121 seconds

Q2: What is that in usable units?

A2: 27,953.8 – 39,135 minutes

A2a: 465.9 – 652.3 hours

A2b: 19 – 27 days

As you might have noticed, the result above does not get us to what I mentioned above (~3-day trip to orbit). The reason for this is twofold. First off – while the anchor for the elevator needs to be in GEO, the actual 'top floor' can be at lower orbits. However, doing so will require a bit more engineering than I am comfortable handling on my own.

Second, the primary limitation on elevators here on the ground, is the time required to stop the elevator. So, an elevator could theoretically go significantly faster, if it had literally hundreds of miles of distance to stop.[7] Yes, we could have calculated the actual amount of time an elevator would have to accelerate and decelerate on a space elevator. However, that is a lot of math that I don't have time for.

Space Train

Let's follow the irritation Nicole Stenson feels, as she realizes travel in space is less cruise ship, and more subway. It certainly is a far cry from travel by car or air, where direction and speed changes are virtually on-demand.

If you are just joining us, what you are about to read is a fictional account of what real humans will likely experience at some point in our second century in space. I realize at least 20% of you immediately tuned out when you saw the previous story involved math, so let's see if you can hold your focus here. WHAT FOLLOWS IS A FICTIONAL STORY...but there is also a real explanation of the tech, science or other principle which makes a story like what you are about to read virtually nonfiction.

So, let's check out what Nicole is going through.

The Fun Part

December 13, 2071
Approximately 5 million kilometers from Mars

"This is absurd" Nicole said, "what do you mean we don't have a choice but to continue on to the asteroid?"

"Just that ma'am," replied the somewhat cowed first officer of the space liner *WindStar*. "We are relying on the gravity of Mars to slow us as we approach and have only enough fuel storage for emergency maneuvers en-route."

"But James...aren't you the one who gave us the announcement that we will be unable to land on the resort because of the...what was it you called it? A bacteriological contamination of the transit tubes?"

"Yes ma'am" James replied, "I realize the inconvenience this is causing you and the other passengers, and we have already sent a message to corporate back on Earth, and will let you know immediately what remedy they have."

"Inconvenience is having to shit in a zero-G James…Inconvenience is having a water ration for showering…Inconvenience is having exactly NO variation in the planned menu." Nicole was on the verge of losing what vestiges of control she had. Despite growing up in a family where outbursts were anathema, she was on the verge of breaking. "James… traveling three months with 20 strangers, my future ex-husband, and my aging dog, only to find out we did it all for nothing is NOT inconvenience, it is a ****ing crisis."

"This happening on a cruise ship on Earth is understandable, but there we just divert to another port. Are you suggesting the two million dollars I paid for this trip wasn't enough for a little extra gas? This is starting to feel more like a subway trip rather than a solar cruise."

"That is actually a good analogy Mrs. Stenson," James replied. "Within a few days of setting out from Earth orbit we were basically committed to this trajectory. By the end of the first week, we had already burned nearly 40% of our fuel, and our emergency plan consisted of a slingshot maneuver around Mars for a return home."

Nicole deflated a little at that explanation. Intellectually, she knew he was right…but instinctively she felt there must be another way. They were in space after all…how easily did ships alter their vector in the movies? As Nicole returned to her cabin, she figured if she wasn't going to get a chance to lay on the virtual beaches of the Martian resort, she could afford to take a night eating and drinking as much as she wanted. What use was a beach body up here anyway?

The Real Deal

Six weeks after launching from Cape Canaveral, an Israeli-built probe, funded through private donations, arrived in orbit around the moon. The Beresheet spacecraft carried enough fuel for a single maneuver to steer it into orbit around the moon. If the probe misfired, Israeli officials said the spacecraft would have continued on into deep space, bringing the mission to an end.[8]

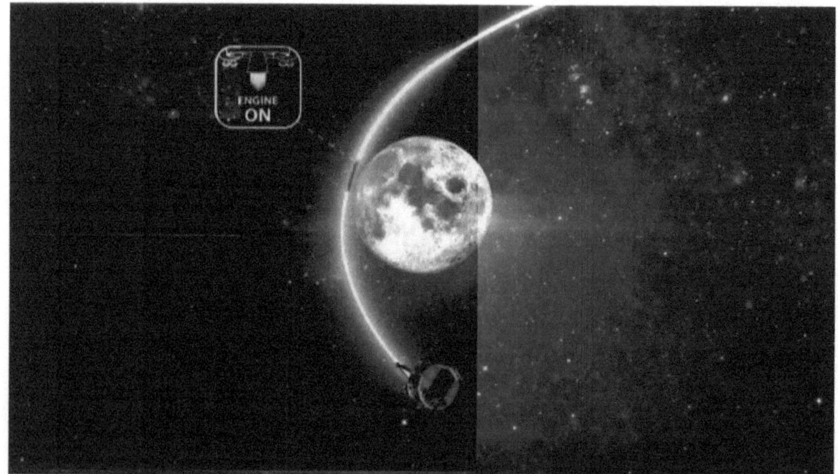

Figure 1: *The Beresheet spacecraft's six-minute deceleration burn steered the probe into orbit around the moon. Credit: SpaceI*

While this was the first privately funded ship to enter lunar orbit, the mechanics of how it decelerated are not unique to any spacecraft.

On earth, we have the delightful benefit of friction[9] to slow anything we happen to be traveling in. This allows us to have our vehicles 'push' their way along. When it comes time to stop, the friction of the ground, brakes, air, or water is used to slow the vehicle, rather than reversing the engine's output (yes, most oceangoing ships do have the ability to do this, because 'braking' at sea is tricky).

In space there is no friction...at least none that will have a measurable impact on the velocity of an object. So, in order to stop, you have two options.[10] First, get yourself captured by a gravity well, like the Beresheet probe did. This involves adjusting your vector and speed slightly, and then entering what is called a degrading, elliptical orbit. This is just a fancy way of saying the orbit looks like a spiral oval.[11]

The second option is to reverse your thrust. This could be accomplished, by having an engine output on either side of your ship.[12] However, an easier way would be using small thrusters positioned along the hull to flip the ship around, until the main engine is facing your direction of travel.[13] This is the space equivalent of pulling the 'e-brake', spinning around, and gunning your car's engine.

This second option requires a ship to carry sufficient fuel to both get up to speed AND decelerate.[14] This would have the practical result of the ship going slower to conserve the amount of fuel needed to brake.

But what about changing direction?

This explanation has largely focused on stopping at your destination, rather than changing direction. This is purposeful, because the principle is the same.

In space, changing direction means changing your velocity, which requires some sort of energy expenditure. Therefore, instead of being able to rely on the friction of air over wings, or the ground on your tires, in space there is no such option. You either use your engine[15] or a gravity well.[16]

These are not great options. However, we chose to use trains and subways to get around here on Earth, because those were faster and more convenient ways to haul people and cargo than the alternatives. It looks like until we get some sweet new tech to take us around the stars without the need for limited fuel, we are stuck riding space subways.

14

Buzz's Cycler

Getting to Mars is a long trip. Right now, most of our plans to get there involve one-way rockets. What if there was an easier way? Buzz Aldrin proposed a better way, that would allow us to ride to Mars in style, for a fraction of the cost of one-way rockets.[17] Let's check out Ben, as he rides the first of these 'cyclers'.

The Fun Part

November 23, 2073
6 million miles from Earth

This was the best use of Marriott points Ben had ever imagined. Granted, it was about 30 million points for the trip. But still it was a good deal. Being in the Aldrin Signature Hotel for the first time brought chills down his spine. Not because of its size...it wasn't much bigger than a small-town motel. It was because it was travelling somewhere between 20,000 and 30,000 miles an hour. He could check his tablet to see the exact speed and position, but it was almost time for the symphony.

As he left the symphony, Ben reflected that this wasn't the nicest hotel he had ever been in. But as far as 'all inclusive' places went, this was pretty good. The food seemed fresh and he wasn't sure he wanted to think about where the water came from. However, there was regular entertainment and plenty of space to mingle with the other guests.

Was it just him, or were most of the women in the hotel really attractive? Why were all of these really hot women headed to Mars? Well, now that he really looked around, there did seem to be more than an average amount of really well-off, and ostensibly powerful men on the trip as well. It wasn't quite the image that Ben had in his head for people who would immigrate to another planet. After all, only a handful of these people were tourists like him. They were leaving Earth, probably for good.

It was weird, he thought. He would think about that later. For now, it was time to go talk to Sophia at the bar.

The Real Deal

A Mars cycler is a permanently orbiting vehicle, with a path that alternately brings it near Earth and Mars. Once a cycler has been accelerated into orbit, it continues on its own momentum, going back and forth between the two planets. It only requires propellant for occasional course adjustments. This cycler concept could also be applied to the moon.[18] It was actually initially conceived as a way to get there.[19]

A one-way trip between Earth and Mars involves five to eight months of space travel.[20] Therefore, a large and well-equipped Mars cycler would offer space explorers, and possibly even space tourists, better accommodations for these long journeys. Smaller spacecraft would ferry travelers between the planets and the cycler.

Buzz Aldrin...yes that Buzz...first proposed a Mars cycler in 1985.[21] The pair of Mars cycler vehicles could provide regular transport between Earth and Mars. While the astronauts traveling to the Moon could do so in spacecraft with a relatively small amount of habitable space, a mission to Mars would require something much larger.

The orbit of the cyclers would be offset (because the orbits of the Earth and Mars don't line up), so that one cycler has a fast trip from Earth to Mars, and a slow return. The other would be the opposite.[22] This would create a system that Aldrin likened to an escalator, where you just transfer between the 'up' and 'down' escalators to go between floors.[23]

I would like to pretend that the image below will help depict this...but I don't know that it will.

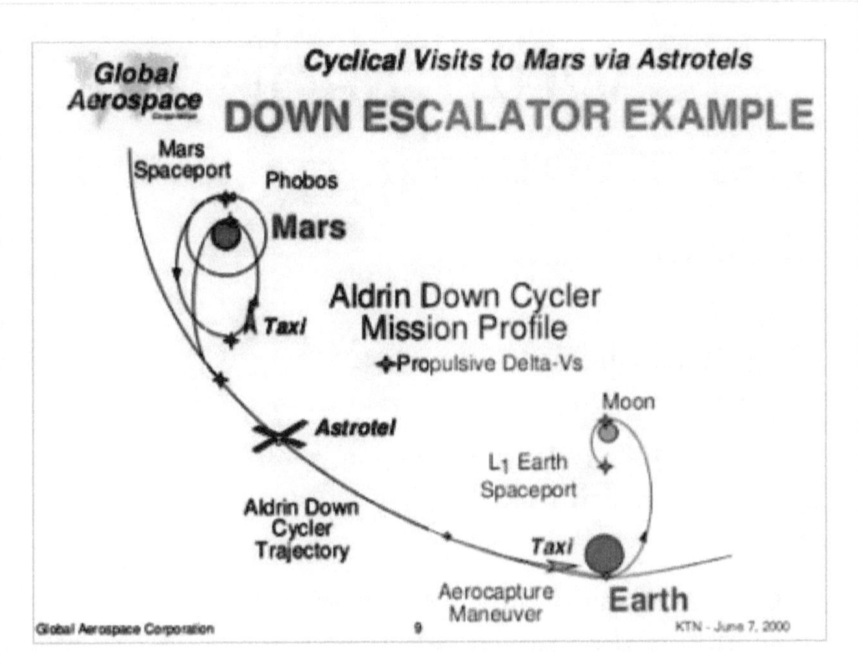

If you think about it, this is a really sweet system. You don't need much fuel to get to/from these cyclers, which can be basically huge orbital hotels.[24] There can be any number of them going between the two planets, cutting down on the wait time to begin your trip.[25]

WHAT ABOUT EATING?

Space Food

—- Begin transmission —*June 4, 2059*
From: Isabella

To: Christof and Aische Connecting.....
 Network Connection established.... Speed: 19mbs
Latency: 9s

Backup routing available..... Receiver online.....
—-Warning: all data sent to and from the European Continent is subject to *General Data Protection Regulation* revision 3 dated March 2046.

Text only record for station archive.....
GDPR compliance audit required.....
Hey Mom and Dad! Thanks for the birthday note yesterday. Things have been really busy up here recently, but it was really nice of you to send that along. I know you that you are still worried about me being up here, but I have never felt better.

—-Receiver response record blocked per GDPR—-

Well, yeah I guess I do look a little worn. Today could have been worse. I mean, don't get me wrong. I really enjoyed my birthday. It would have been better with you all up here. However, all things considered, it was really good. And no mom, I didn't find a nice German boy up here. Isn't that what you always ask me, after I go to a party?

—-Receiver response record blocked per GDPR—-

Hahahaha fair point dad., I guess finding one up here would make it more likely that I did stay up here.

Anyhoo, enough about my love life. I forgot to update you all last week on how far we have come with the surrogate incubation system.

—-Receiver response record blocked per GDPR—-

Sorry, Dad. I know you don't like me calling it surrogate but that is the actual name. You know we use the surrogate bacteria to generate carbohydrate and protein-based substances. I suppose it would be easier for me to just call them the food processors.

—-Receiver response record blocked per GDPR—-

Hey mom, remember that I'll signal when my message is done, so you don't have to worry about missing anything. Just watch the corner of the screen. When it goes green, then I have stopped transmitting.

—-Receiver response record blocked per GDPR—-

Yes, I know this was easier when I lived in Berlin. However, to be fair, the view from here is way better than my old apartment, and I still don't have blinds

—-Receiver response record blocked per GDPR—-

OK, back to the food processors. Dad, they are so much more than that! We are using the proteins to repair our suits up here. I think there is a possibility that we will end up being able to actually grow custom foods that actually taste good with this process!

—-Receiver response record blocked per GDPR—-

Haha, you are right. No one is going to be a food tourist up here anytime soon.

—-Receiver response record blocked per GDPR—-

Thanks, mom. I don't mind dad's interruptions. Usually in our conversations, I give him a hard time for talking in German to try and fool the spies he is sure are watching me. So, I can tease him right back.

—-Receiver response record blocked per GDPR—-

So, back to these processors. You remember when I stopped doing that artificial intelligence ethics work right? It was so weird to get a job working on this project to use bacteria to grow carbohydrates from carbon dioxide. I figured that it would just be a stopover job, until I found something I could really get into.

—-Receiver response record blocked per GDPR—-

Exactly, I didn't think I would find something I liked this much, based on my background in political science and that time working on AI. But man, this is some fascinating stuff. We have managed to cross the 10g/liter production threshold per week, which is huge. We are already able to generate about 100 liters of bacteria a week, and will keep these bacteria in the water these cargo ships are moving around out here.
It allows people to have both radiation shielding and food! It is so awesome.

—-Receiver response record blocked per GDPR—-

It is true that these bacteria are resistant to radiation. That allows them to live in the radiation shield. However, we are making sure these and all subsequent strains remain non-threatening to humans.

—-Receiver response record blocked per GDPR—-

Hahaha, no it doesn't work like that mom. I think that you have been watching too many old movies. These bacteria actually never leave the radiation shield. The product we get out is carefully treated to make sure it isn't contaminated. All ships still keep at least a week's supply of emergency food on hand, just in case. The worst-case scenario would be that the bacteria all die. The antibiotics that are available in any ship's medical kit would be enough to help anyone who happened to get an infection.

—-Receiver response record blocked per GDPR—-

Oh, OK. Sorry, I guess I was probably getting a little boring there. Yes, I am really enjoying my job up here. Hopefully, I'll still be able to be back for Oktoberfest. I'll keep you updated.

—— End transmission —-

—- GDPR record audit complete —

The Real Deal

Seems a tad far-fetched – right, to have bacteria grow food in space.[26] Well, right now, NASA is sponsoring research into just that (well, technically it is research into bacteria growing silk).[27] So far, the study is showing pretty good promise. It is a few decades away from being able to supply even a single person with enough nutrition to sustain them, but it will likely emerge as a key component in future long-distance space missions, where weight and space will be at a premium.

The ability to engineer bacteria to be both hearty enough to survive the microgravity[28] and radiation in space, and prolific enough to make enough protein, carbohydrates, spider silk, or whatever else will no doubt be tricky business. I don't envy the people who have to taste test any food that comes from this. However, at the end of the day, the ability to self-generate food will be a huge boon to any space-based community.

Iron Chef Orbital Edition

If you are a normal human, you have the ability to cook your food. You may not have much food to cook, but you do have at least a rudimentary ability to prepare and eat food. What about in space? How do our current space peeps cook? How are we going to cook, once we head out into the black? Let's explore a possible future.

The Fun Part

June 10, 2029
International Space Station, Low Earth Orbit

"Well, this is going to be tricky," Jackie thought. How exactly am I going to mix this in here?

As Jackie looked at the pan holding the chocolate, butter and sugar, she realized that making the cake from scratch may not have been the best idea. Especially since the pan wasn't even really 'holding' the ingredients…it was just sitting there, while the ingredients sort of floated above it.

Well, I guess I could just throw the ingredients in the convection oven for about five minutes, and then try and blend it all together somehow. It wasn't like anyone was going to have another cake to compare this to up here.

After the quick melting of the ingredients, Jackie tries to put the (mostly) unmixed, mix into her blender. She quickly learned that the blender only actually mixes things when there is some sort of force pulling the ingredients towards the blades at the bottom.

"Maybe" she thought, "I should just give up on trying to bake a cake, and just work on dinner."

"Or" she muttered, looking at the chicken cordon bleu recipe she brought from home, "I should just focus on making the de-hydrated food edible."

The Real Deal

While Jackie's cooking experience above may have seemed a little silly, there is, unfortunately, more truth to it than you might want to believe.

How we do it now

So long as we are being honest with each other, there isn't much 'cooking' happening up in space. A few years back, chef Heston Blumenthal[29][30] collaborated with a British astronaut to design a better menu for the astronauts.[31] However, according to Blumenthal there was nothing but obstacles all the way. First among the obstacles, were the tools available.

The two cooking related tools currently on the station are the index finger (for hooking ring-pulls) and a pair of scissors (for opening pouches). Otherwise, meals are heated in a compact convection oven about the size of a briefcase.

"Space food is very, very controlled," says Blumenthal. "All this red tape! You've got to get [each dish] past the UK Space Agency. Then the European Space Agency. Then there was NASA. The people in these organizations who are involved with food, they're... they're engineers." Not chefs, he means. Not gourmands. "They consider food to be fuel. A lot of freeze-dried stuff. Tins. Pouches. I was shown these packets from the ISS with big Russian writing on the front. Stuff that looked like you shouldn't be putting it in your body at *all*."

This "two-year project was challenged as much by Blumenthal's personal resistance to dullness, as by any agency's red tape. Space food has to be dull, to an extent, because neither pizzazz nor subtlety can be absolutely trusted up there"…yet. Early ideas included "sending up smells in spray cans, to be spritzed around the ISS while eating. It was better but NASA was not into that."

What NASA had been into, was function over form. As recently as the 1990s it was being openly proposed that valuable storage room be saved on board spacecrafts by finding ways to reconfigure astronauts' waste. One strategy meeting at NASA, apparently, ended with a pilot saying: "We are not eating shit burgers."

Since the 1990s, few astronauts have bothered experimenting with cooking in space. A notable exception was Sandra Magnus,[32] the flight engineer for ISS Expedition 18.

Magnus undertook a number of cooking experiments,[33] aided by cutting boards and bowls that she anchored down with copious amounts of duct tape. Perhaps the most impressive was her re-purposing of the ISS food warmer to make roasted garlic and onions, an operation she achieved by using foil packets saved from previous heat-and-serve meals, to run vegetables through the food warmer again and again for hours at a time.[34]

What other limitations are there?

Taste and preparation are not the only two factors you have to think of, when approaching the problem of cooking on space. The physics of cooking itself are altered by the conditions of space. Our cooking techniques are honed at Earth's gravity and, when we take those techniques to other gravities, the results can be surprising.

Jean Hunter, a professor at Cornell, who studies how food and cooking work in space, explained how even something as simple as making a hard-boiled egg becomes difficult, without Earth's gravity to aid in the boiling process.

"The problem with cooking in space is that there's no gravity," Hunter told an interviewer. "If you wanted to boil an egg, for instance, the vapor and the liquid won't separate. Rather than the water boiling like on earth, where the bubbles rise to the top and release steam, what you'd get is more like a can of soda flowing over. It would be very difficult to cook with it."

I should take a break here to say no…cooking in space is not like cooking at high altitudes…other than the fact that you are literally high in altitude. The reason you have to adjust your recipes from sea level to Denver is because the air pressure is different which changes the chemical reaction of the ingredients while they bake/heat/burn.[35] In space, assuming you keep the interior pressurized,[36] you will not have the issue of air causing your baking to mess up.

Historical Analogues

Explorers and travelers throughout history have had to develop methods for preserving food and carrying enough food for their journeys. This problem was especially difficult during the time when people made long sea voyages on sailing ships. Great explorers like Columbus, Magellan and Cook carried dried foods and foods preserved in salt and brine.[37]

However, by the time the British Empire ruled the seas, things had gotten quite a bit better.[38] British warships were some of the best equipped and provisioned ships during the age of sail. For instance, packet boats from England brought fresh food to the Channel Fleet and some captains kept gardens ashore along their assigned cruising lanes to supply the crews with vegetables.[39]

The officers were the best off, but even the common sailors had their food prepared on a main stove or baked/roasted in one of the ship's smaller ovens (like the ones shown above).

While other nation's naval food quality and quantity varied, the general trend was moving from the dried or pickled food of the 1400s to the kitchen prepared food of the 17-1800s.

Likely next steps

OK, so great. The sailors in the good old days went from eating hard, dried food to having a chef prepare their meals in a kitchen, often using fresh(ish) food. What does that mean for us?

I gave the examples of the age of sail, because it is likely a similar progression to what we will move along. Right now, the height of 'kitchenware' in space is a convection oven, which while useful, is not exactly the only cooking implement that spacers will want.

In the next 10-15 years, expect a contained stove type system to be introduced whereby an induction-type heating element is used to cook food in a pan.[40] A system which uses forced air to 'push' the food into the pan may be one solution to keep the food from floating away. Another is to have the food packaged in a magnetic container and to just remove the lid for cooking.

Simultaneously, I expect the increase in wealthy space tourists to spur innovation in Earth to orbit food transit, where fresh foods can be more readily transported into orbit. This will open up new possibilities for cooking, even with a limited ability to heat the food. And realistically the ability to move pepperoni, cheese and pizza dough into space are three of the four things keeping me from living in space. The other one is money...oh...or is it a bed...I can never remember which one of those would be more important.

WHAT ABOUT THE BORING STUFF?

The View...

The view really was worth every penny they had paid for it, Kelsey thought as she looked out of the small window in her new home. Although, she thought, the pictures really did make it look bigger.

Kelsey and Adam had purchased their new vacation home, a re-purposed upper stage from Blue Origin's New Glenn rocket, about six months ago. It was right before it had launched into orbit in the fall of 2049.

Their home had been maneuvered into a stable low earth orbit, and the renovation teams had worked quickly – all things considered. Their home was positioned with about 75 other boosters forming a sort of three-dimensional neighborhood, although this neighborhood was held together by tether cables rather than roads. Glennville was what some were trying to name the neighborhood. Kelsey figured it was as good a name as any other, although she was sure there would be someone at the association meeting who would object when it came time to vote.

Speaking of that association meeting, Adam was supposed to talk with the current president at his weekly poker match. Kelsey instinctively looked out the window again, but this time she was searching for movement between the habitats. Sure enough, there was Adam, attached to one of the tethers, being slowly hauled towards the roundabout in the center, when he would latch on to their tether, and begin the process of cycling the airlock.

It couldn't be more than 100 meters between the habitats, but Kelsey knew from frustratingly common experience that the distance she could cover Earthside in a matter of seconds, took more than 15 minutes out here. Adam had at least another nine minutes before making it back, and six to cycle the airlock and get his suit off, assuming he hadn't had too much to drink. It just enough time to finish her email to her team back at the company's HQ on earth.

The Real Deal

It is not quite the science fiction story you would expect for someone writing about the future of space. However, that's the point. The future will be just as mundane as right now is. It will just have a different view.

As of early 2019, the Jeff Bezos' backed Blue Origin company is exploring keeping the upper booster stages of their New Glenn rockets in orbit, instead of de-orbiting them.[41] The thought is that these boosters are basically ready-made 'mobile homes,' that have been given a little retrofitting in orbit.

This solves one of the most significant problems with building orbital habitats now. It costs a lot to haul a structure up to orbit. Moving smaller parts like airlocks, couches and heaters is much more manageable. As a bonus, Blue Origin will be selling space on the rockets they launch,[42] so keeping the boosters in space is basically a free way to make some side money for them, and a great way for you to get one of those 'fixer uppers' that you keep hearing about on HGTV.[43]

What Time Is It?

So, there is an easy answer for this on Earth now that us fancy monkeys[44] have figured out that it is efficient to make zones for time. However, that system makes no sense once we are on the moon, in orbit, on an asteroid, or on Mars. How the 'f' do we figure out what time zone Jupiter is in? Well, good news for you. There isn't any good news. Unless we have some sort of human-universal time, we are going back to the time when every town set their own clock.

The Fun Part

3 February 2031
The Kanye Low Orbital House of Kool (Formerly known as the International Space Station) 2130 Shipboard time / 1215 PST

Hey boss, your wife Addison just called. She seemed a tad agitated. What do you mean John? God, it feels like I just got to sleep.
You did sir. I'll send you the recording. The short version is that she just went into labor, and wants you back.

Damn John. OK, sorry that you had to take that call. Just send me the recording and I'll call her back after I listen to it. Can you ask Emily to spool up the lander? I'll head towards the airlock as soon as I change.
——Recording Begins——Caller: Addison
Location: San Francisco

What do you mean he isn't available? Is he in a meeting with the pope? I told him that unless he is dying, he better be available for my calls. Listen very closely John. I recognize that you are trying to protect your boss. However, I am going into labor with his kid here. What exactly is so important, that he isn't there in the office?

Wait, what do you mean he is in his room sleeping? It is lunchtime here.

Shipboard time? What is shipboard time? And why did he forward his calls to you? How come you are up?

John, you are not helping him at all by stalling like this. Just go get my husband and let him know he better be on the first shuttle back down here or I will remove his trachea when he gets back.

——Recording Ends——-

John, are you still there in the comm room? Yeah boss, what's up?
Seriously dude, I'm sorry about that. Addison is normally such a nice person. I have no idea what is going on with her.

Oh, no worries sir. I have two kids. This is your first, isn't it? Yeah.
Oh well, labor is a real bitch, anything that a woman says while in labor, should never be counted against them. I certainly don't.

Well, thanks John. I'll probably be groundside for a couple weeks. Shoot me a note telling me what your favorite scotch is, and I'll bring it up next time I come.

The Real Deal

We all have time zones. Realistically, there aren't a lot of us around who were alive when they weren't in widespread use. However, they are a fairly recent innovation. Until the late 1800s, there was no such thing as time 'zones' in any standardized way. Prior to this time, each town set their own time, according to their own sundial, church clock tower, or some other timekeeping device.

The advent of railroads made this system unworkable. However, there wasn't an initial consensus about how time should be kept. The International Meridian Conference in 1884 was the first step in a process that ultimately created the world's time zones as we know them. However, it took until 1920 for most countries in the world to fully get on board. Even now, countries like the U.S. continue to tinker with things like daylight savings time (the most recent change to DST was in 2007).[45]

Cool, but what about space?

That is a great question. However, this whole book is about space, so maybe you should just be a little more patient.

The whole basis for our time zones, as established by the International Meridian Conference, is degrees of longitude. For those of you who slept through geography class (probably every single one of you American readers), those are the lines which run up and down on the globe. Each time zone is roughly 15 degrees of longitude wide...or as wide as 1,100 miles at the equater.[46]

Yes, I am getting to how this won't work in space. Although if you need to be told that there is no longitude in space, then you really shouldn't have slept through that geography class.

It is true that there is no longitude in space. Therefore, our current time zone system won't work. We could decide that stations in geostationary orbits could share time zones with the part of the earth they permanently orbit over. This would probably be a good system. However, with our only space stations having been in lower orbits, we have not yet had that option (the international space station currently orbits the globe about once every 90 minutes[47]).

We also can't use the sun to determine when it is 'day,' because most places we will be in space are not going to have a 24-hour day/night cycle. For instance, the moon has a 'day' equivalent of 29.5 Earth days, while Mars' equivalent is about 25 hours.[48]

One of the most likely solutions is to have all stations, habitats and other locales in a given area (low earth orbit | Lagrange points | moon ...etc.) in the same time 'zone'. The other probable solution is the permanent adoption of UTC as 'space' time (this is what the International Space Station does now).[49]

The second solution would be the more efficient one. However, remember that we are fancy monkeys, who don't like being the same as the other monkeys. Therefore, people are going to want to have their own special times.[50]

Any way we do it, time in space is going to be arbitrary, compared to what we are now used to. Fortunately, we all are used to carrying around devices that automatically update to the current times. That will at least make it easy to know what arbitrary time it is, wherever we are.[51]

Can We Build That Here?

How long will we be building space systems that are: (a) way bigger than they need to be and (b) completely customized for a single purpose? For comparison, we are still operating on the mainframe computer model, where computers 'had' to be the size of rooms. That didn't last for computers. It is just a matter of time before we transition to the desktop computer model of being able to just plug and play with dozens of different modules. One company is working on just that future. Let's see what Deb is doing, before we get there.

The Fun Part

July 4, 2031
Geostationary Orbit somewhere over the Middle East

"I suppose that I should be grateful, at least I have a job safe from the recession Deb thought, as she clipped into the school-bus sized Defense Department satellite. It wasn't like she disliked being on the Space Force repair detail this quarter. In fact, this was one of the better temporary duties she had been on, since joining the military eight years ago.

The thing was, there was no reason for her to be here. It was true that the Space Force's newest communication satellite was definitely broken and needed to be repaired. However, SHE didn't need to be repairing it. There was no reason. Well, maybe there were reasons, although they were not good ones why satellites needed to be uber-specialized multi-billion-dollar systems. It is not true anymore.

That would be like telling an iPad user that it made more sense for them to buy a house-sized mainframe for their next computer. Nothing else in life was so specialized, that only a small handful were ever made. This is especially true for systems that cost billions of dollars and did only one job.

This communication satellite, she thought, just took a signal from either the ground or another satellite, and rebroadcast it in the opposite direction. It was a little more than that but why was it so big?

"More to the point, why was it so hard to fix anything in here," Deb thought as she opened up the first external panel. Today's spacewalk was scheduled to last four hours. There were five more planned as backups, in case she didn't fix the problem right now. If she was honest with herself, it was going to take at least two of those backups. She was going to be lucky to just finish examining the receiver here, before she needed to head back.

As she worked, Deb started reminiscing about the article she wrote for her school newspaper on the self- assembling, modular space station that had been launched, when she was a senior nearly nine years ago. They were the sort of systems that she thought she would be working with, when she joined Space Force. However, it looked like at least the next generation of military satellites were going to be these monstrosities. They were not at all like the modular, upgradeable systems that most companies were starting to use.

Now that we can manufacture systems up here in orbit, there isn't a need for systems that have to survive the trip into orbit. At least there wasn't the gawd awful smell the swap like back home in Florida...there were worse places to work.

The Real Deal

I interviewed the CEO of Tether's Unlimited Inc (TUI),[52] Dr. Robert Hoyt.[53] In addition to being a good enough person to actually take my call, he seems to be one of the few people who is more optimistic about the future of space than I am. We'll get into more about that in this chapter However, he believes that before he retires, there will be a self-sustaining economy in space.

Self-Assembling Stations?

What Deb was talking about in our story? A self-assembling, modular space station seems like the sci-fi of the 1960s. However, Dr. Hoyt is confident that his company will be able to test this concept in the near future. He talked to me about TUI's Constructable™ Platform concept, which is part of a DARPA[54] project on which TUI and Space Systems Loral (SSL)[55] are collaborating.[56]

TUI's Constructable™ Platform is simultaneously as simple as it sounds, and really hard. The general plan is to stitch together Microsat-sized (or smaller) modules into a larger structure that shares power, communications, station-keeping, and other key services. While the DARPA project is focused on traditional security missions, the platform would have the ability to grow far beyond whatever its original mandate was.[57]

Figure 2 *An image of what an early Constructable™ Platform would look like. Credit: Tethers Unlimited*

An image of what an early Constructable™ Platform would look like.[58]

Once in orbit (which Dr Hoyt believed could happen by the middle of the 2020s), the Constructable™ Platform would function as a sort of Lego style base for 'tenants' to build off of. If a company's module was compatible with TUI's platform, it could be integrated. This gives companies, universities and countries the inexpensive option of slowly upgrading their in-orbit infrastructure without wasting resources.[59]

Space Spiders

Figure 3: Artist rendering of a SpiderFab. Credit Tethers Unlimited

[60]The Constructable™ Platform is just one of the on-orbit manufacturing concepts that TUI is working on. Another is the "SpiderFab" system (shown here).[61] It is designed to build multi-kilometer long structures in orbit through what is essentially 3D printing. The initial test of **SpiderFab** will be to work in tandem with another TUI tool the **Trusselator** to make long trusses in orbit.

As the truss is built, solar cells (or antenna material, or solar sails) slowly unroll along the top of the truss (see below).

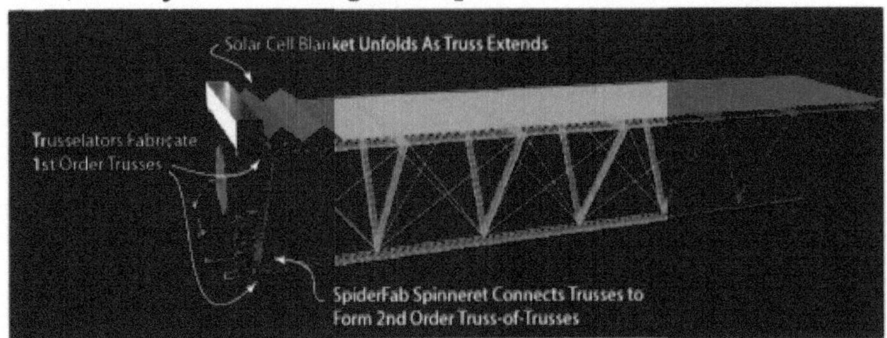

So What's The Holdup?

Because I like knowing what is preventing the future from being here now, I asked Dr. Hoyt how fast he could get the Constructable Platform into orbit, assuming that money was no object (my actual question was if he got a billion dollars in no strings attached venture capital overnight). He said with that money, he is pretty sure they could get a demonstrator platform into GEO in two to three years and have it fully operational within five years.

TUI ultimately wants to eliminate the man in the loop for all their on-orbit manufacturing systems. Dr. Hoyt sees a future where someone in the group just sends a design file to an on-orbit system and it is able to build that entirely autonomously. If you don't find that exciting, then you need your head examined.

More than just a Gateway

"Skeptics, that's what you all are." "Not a god-damned believer among you."

Those were my last words to my family, before I boarded the New Glenn rocket for my trip to the Gateway. Oh damn, that wasn't how I wanted this video to start.

OK, let's try that again. My name is Kevin. I am the first specialist assigned to Gateway after it was completed last year.

Specialist in what? I'm the first computer scientist up here.

There have been others with computer science degrees who were astronauts. However, I am no astronaut. I'm just the lucky nerd, who was in the right place at the right time. I started the artificial intelligence company BlackNet. We made great progress in applying machine learning theories to orbital surveillance and cislunar exploration.

 It means that we are using some fairly complicated custom AI processors slaved to a suite of hyper-spectrum sensors to find resources for mining companies to exploit. It's all pretty easy. However, as my ex says, no one cares about the details, except maybe me.

Why am I up here? I'm still asking myself that. The official reason is that the equipment didn't work how we advertised. It was cheaper for me to come fix it, than it was to send it back to earth and repair/replace it.

It sounds crazy, right!? With the cost of getting to Earth-Moon Lagrange Point 1 now down to $3k a kilogram, I'm not THAT expensive to get up here. It only costs about $250k. It is a business expense that I can deduct from BlackNet's taxes. Therefore, after taxes it will only cost… okay, maybe not all math is easy in my head.

I'm up here to service the processors, because they have been reacting weirdly to the radiation and the intermittent loss of connectivity to earth.

It hasn't been since early 2019, that anyone really thought of this. That was when Hewlett Packard finished their experiment with a couple server racks in space. They pioneered the algorithms we all have been using to make cheaper, less shielded electronics for at least the past ten years.

I am sorry, this is supposed to be a promo video for stupid investors. Jamie, could you make sure that you cut all these parts?

Let's go back to the HP experiment. They found that electronics outside of the Earth's magnetosphere could be taught to weather solar storms and radiation, by entering a hybrid 'safe mode'. We all thought that the hybrid mode would work out here at the Gateway. It turns out that we failed to account for what happens here, where the gravity fields of the moon and Earth cancel each other out.

Although I am not an astrophysicist, what I understand is that in Earth orbit, and even to a degree in lunar orbit, there is something that disrupts the radiation just a bit, and that makes it easier on electronics.

I know I haven't gotten to why I am on a three-month, half-a-million-dollar vacation out here. Why couldn't I just troubleshoot this from dirtside? We tried that, and no offense to these professional astronauts, but they really are bad at what can only be described as surgery on these computers. After NASA paid about two billion dollars to move this system way up here and plans to begin selling the data it finds to the exploratory mining consortium, there was pressure to get this system up and working ASAP.

That is the reason why I am here. I realize I still haven't gotten to the point of this fundraising video. I just hope that it has been edited to make me sound professional.

Even though the system up here shut down after the first solar storm, the sensors were still working. There just wasn't any processing of the data. The government technically hasn't 'accepted' this system. They told me that if I couldn't fix it, then they weren't going to. That tells me that I still own the system and the data it collected. The data shows there are tons of rare minerals below what appears to be a crash site of an old lunar explorer at 119.1 degrees east longitude and 3.0 degrees north latitude.

Here's my proposition. I'll tell NASA that the system is flawed and must be replaced. It is a process that will take at least a year. I'll offer to split the cost with them, so they don't complain. During that time, we will form a company and begin launching those new Toyota lunar buggies with the mining attachments, as well as a handful of those modular smelting and processing stations to the area around the crash site.

Based on my back of the napkin estimates, those minerals would be worth about $650 billion. Our startup costs would be about $4.5 billion, and it would cost about $150 million per return trip so we would gross at least a billion dollars.

Our year head start means we won't have to pay anything for this data. The government also doesn't get to dole out mining rights, based on a senator's relationship with his secretary. It also means that we would get at least six shuttles back to earth. This would allow us to recoup our setup cost in less than the first year.

Let me know what you think. Please make it quick. I have other investors who I can reach out to, if you aren't interested.

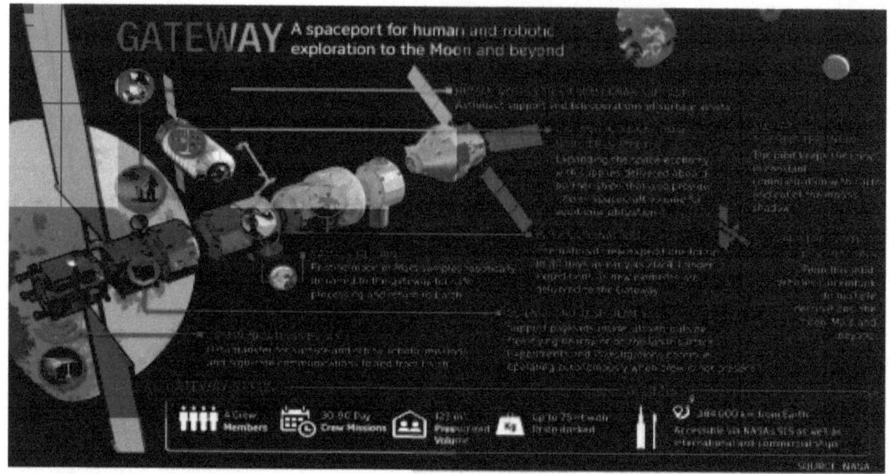

NASA's concept for the Gateway[62]

The Real Deal
Gateway

As NASA sets its sights on returning to the Moon, and preparing for Mars,[63] the agency is developing new opportunities in lunar orbit to provide the foundation for human exploration deeper into the solar system.

Since at least 2018, the agency has been studying an orbital outpost concept in the vicinity of the Moon with U.S. industry and the International Space Station partners. As part of the fiscal year 2019 budget proposal,[64] NASA is planning to build the Gateway in the 2020s.

The platform will consist of at least a power and propulsion element and habitation, logistics and airlock capabilities. While specific technical and mission capabilities as well as partnership opportunities are under consideration, NASA plans to launch elements of the Gateway on the agency's Space Launch System or commercial rockets for assembly in space.[65] It appears increasingly unlikely that the SLS system will be the one that puts all the Gateway in cis-lunar orbit based on design delays.[66] However, with other systems like

Falcon Heavy being able to reach lunar orbit,[67] there are other commercial options for getting the station up and running.

The power and propulsion element will be the initial component. It is targeted to launch in 2022. Using advanced high-power solar electric propulsion, the element will maintain the Gateway's position and can move the Gateway between lunar orbits over its lifetime to maximize science and exploration operations.[68] Habitation modules that are projected to launch in 2024, will further enhance the station's ability to have scientific, exploration, as well as commercial use. It isn't yet clear what NASA's plans to return humans to the moon by 2024 will do to this timeline. However, it is likely that it will be accelerated. Once the habitation module is in place, the crew aboard the Gateway could live and work in what is technically deep space for up to 30 to 60 days at a time.[69]

Moon Mining

In-situ resource utilization (ISRU) is the technical term for identifying, extracting and processing material from the lunar surface and interior and converting it into something useful:[70] oxygen for breathing,[71] electricity, construction materials[72] and even rocket fuel.[73]

There have been discussions of eventually mining and shipping back to Earth the helium-3 locked in the lunar regolith.[74] Helium-3 (a non-radioactive isotope of helium) could be used as fuel for fusion reactors to produce vast amounts of energy at very low environmental cost – although fusion as a power source has not yet been demonstrated,[75] and the volume of extractable helium-3 is unknown.[76] Nonetheless, even as the true costs and benefits of lunar ISRU remain to be seen,[77] there is little reason to think that the considerable current interest[78] in mining the Moon won't continue.

It's worth noting that the moon may not be a particularly suitable destination for mining other valuable metals, such as gold, platinum or rare earth elements.[79] This is because of the process of differentiation,[80] in which relatively heavy materials sink and lighter materials rise, when a planetary body is partially or almost fully molten.

This is basically what goes on, if you shake a test tube filled with sand and water. At first, everything is mixed together, but then the sand eventually separates from the liquid and sinks to the bottom of the tube. Similar to Earth, most of the moon's inventory of heavy and valuable metals are likely deep in the mantle or even the core, where they're essentially impossible to access.

Launch Costs

Since 1980, the cost to move a single kilogram (kg) of 'stuff' (that's the technical measurement used by NASA) to low earth orbit has declined nearly 100-fold.[81] The graph below shows the current trend, as well as a projection of where costs are headed based on the last 40 years.

[82]

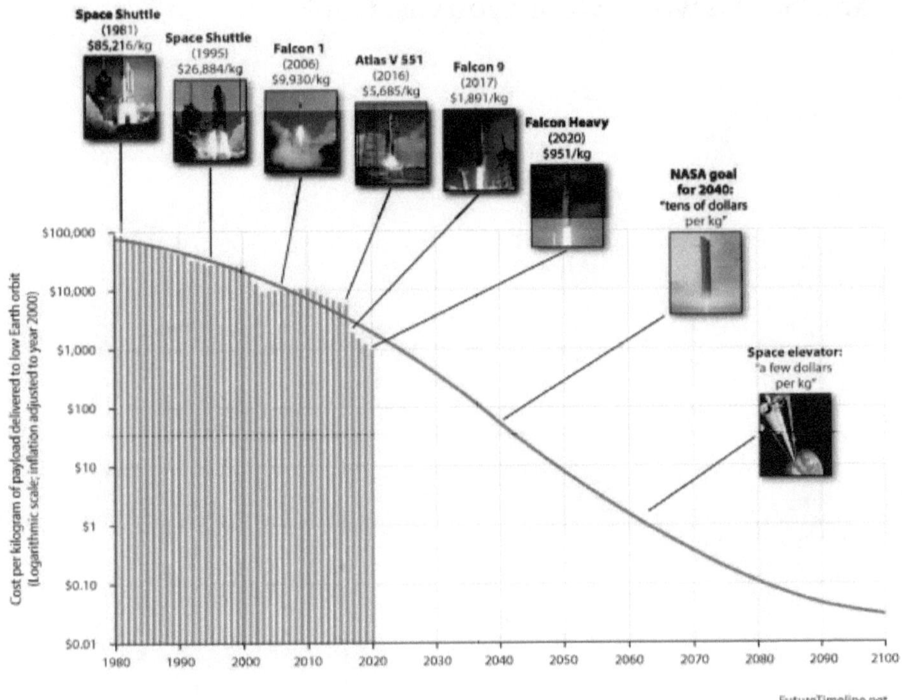

Launch costs to low Earth orbit, 1980-2100") and then following it with

To put this in context. I currently travel to Minneapolis once a month. A one-way ticket from my house to there is $250 (you can find cheaper ones, but my ticket has someone serving me alcohol while the rest of the plane boards). I weigh roughly 80kg, which makes the travel to Minneapolis cost roughly $3/kg.

Based on the current trend-line, we will reach that cost sometime between 2040-2060. Let that sink in for a second. By the middle of this century, it will be as cheap to get to orbit as it will be to fly halfway across the US.

Humans living, working, mining and generally existing in space still seems like science fiction to the majority of the public. However, if you could fly to orbit or Cancun for the same price...where would you vacation?

A Not-So Simple Tool

December 13, 2033
Low Earth Orbit, over Brazil

"Well, today is shot" Katie thought, as she received an alert on her tablet warning her that the micro-meteor shield on her room's external internet receiver was damaged. It wasn't that the repair job itself was hard. It just involved replacing the panel, and hoping that nothing inside was too broken. Considering her access to the internet still seemed good, there was a chance this would be a simple fix. However, anything that punched a hole through the micro-meteor shield had to go somewhere. Everything inside that compartment was more than a little fragile.

It was a four-minute trip from her room to the habitat's main airlock. There was a secondary airlock after you entered the main research complex. However, getting from that airlock to the damaged compartment would literally take around an hour. This job was already going to take the better part of the day.

Katie was a neuroscience researcher at the U.S. Orbital Science and Technology Station, a public-private partnership between about 40 research institutions, the federal government and commercial companies. Its workforce of 78 scientists worked primarily in biological fields, with the big tech companies doing more industrial and technical research at the now commercialized International Space Station.

It wasn't that Katie was bad at fixing things. Anyone who knew her, would say she was actually pretty good at figuring out how to fix mechanical things like this. The problem was gravity. It was always gravity. The suit also didn't make things any easier.

Katie laughed to herself, as she maneuvered into one of the utility suits near the airlock. It had been almost 15 years since she first heard anything about women in space suits. There had been a story about how NASA didn't have enough in the right size for two women to spacewalk at the same time.[83] Times had changed a bit. It was now just about as easy to get a suit made for up here, as it was to get a car. It was not cheap, but easy enough to find.

The trip from home to the small compartment outside Katie's room only took about 40 minutes. In

Katie's opinion, that was some sort of record. She made a mental note to double check, before she put up a plaque on the airlock showing the fastest transit times.

Looking at the receiver compartment, Katie's glee at how much of a winner she was quickly faded. The innards looked like an angry child had punched their tiny fist of fury through the protective panel. From what little Katie could see, including the mess of wires inside and bits of debris floating out from the hole, she knew the entire compartment would need to be replaced. That was going to be a long process. It looked like her instincts were right. Maybe she would talk with her husband about them keeping track of whose instincts were more right. That would be a useful distraction.

The Real Deal

Spacewalks have long been a source of frustration for astronauts because of cumbersome suits,[84] a short air supply,[85] and the fact that you have roughly the same ability to perform manual labor as a small infant, due to the lack of leverage.[86] A spacewalk on April 8, 2019 took about 6.5 hours. It involved two astronauts replacing a battery charger, swapping some wiring and adding a jumper cable to a robotic arm.[87]

What do I mean?

A lever is a simple machine consisting of a rigid bar that rotates about a fixed point, called a fulcrum (it is like your arm, in that case, your elbow is the fulcrum). To move an object with a lever, force is applied to one end of the lever, and the object to be moved (referred to as the resistance or load) is usually located at the other end of the lever, with the fulcrum somewhere between the two. By varying the distances between the force and the fulcrum and between the load and the fulcrum, the amount of effort needed to move the load can be decreased, making the job easier. [88]

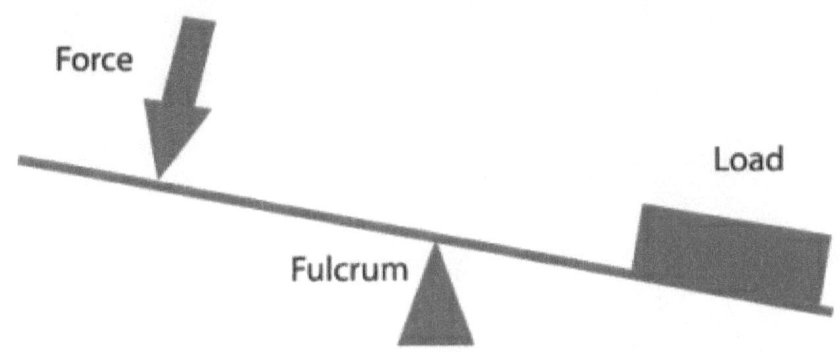

Simple diagram of a fulcrum...it is the triangle thingy in the middle

It seems simple enough. However, in space, there is nothing providing a counterforce to any force you apply to a lever. Imagine that you are floating in a pool. How easy is it for you to play tug of war with someone on the side of the pool? It is pretty hard, right? You have nothing to brace yourself against, while you pull. A similar problem occurs when you try and lift something underwater, without standing on the ground or kicking your legs.

In orbit you have to contend with the difficulty of trying to find something to brace yourself against for every action. You must also do that, while in a suit that is doing its best to keep you alive. However, you do not have any of your normal flexibility or dexterity.[89]

In this type of environment, even just putting up a new curtain rod would be a significant emotional event, not to mention doing something that requires you to do more detailed work, like wiring...or pressure washing the exterior.

War On Titan

Alright, today we get to have some fun. Let's journey far out into the future and our solar system to check in on Lord Cornwall. Along the way, we will explore the difficulties with military communications, and why transit hubs are a big deal.

The Fun Part

December 30, 2099 Titan, Jupiter Orbit

It was just past mid-day, local time, when Lord Cornwall, Governor General of Her Majesty's Colony on Ganymede was interrupted by a blaring alarm on his phone. "Stupid security drills" he mutters as he pulls it out, blanching as he reads the alert. TO ALL HER MAJESTY'S GOVERNMENT PERSONNEL ON TITAN, THE EMPIRE OF BRITTANIA IS NOW AT WAR WITH THE REPUBLIC OF JAPAN AND OCEANIA. UNTIL FURTHER NOTICE, SECURITY PROTOCOL EPSILON IS IN EFFECT.

The governor's aide bursts through the door to his office, as the governor gets to the end of the message. "Sir, I have alerted our Royal Marine commander, and your cabinet. They are en route." The governor sighs, and nods. "James, do you realize how bad this is for us?"

"What do you mean sir? We are at least three days travel from the nearest Japanese outpost, and ten from their closest fleet." Lord Cornwall shakes his head, "No, no, no, James, that isn't it at all. Security Protocol Epsilon will cripple our economy." "You know we are the transit hub between the Belt, and commercial exploration fleets further out."

"Of course, sir" James replies, "most of our income is as a waystation for those fleets. Why would that stop?" "Because James, under Epsilon we are no longer permitted to use wireless communication. With how widely the radio and laser signals attenuate as they reach the inner or outer system, there is

literally no way to know who else could intercept the signals."
"Epsilon orders all British facilities to switch exclusively to
in-person or hardwired communications...which as you probably just figured out, means we should probably have main-
tained our courier boats better."

The Real Deal

OK OK, you are right, this is way more fictional than most of
my stories. This is true. How could I pass up the chance to
write about a governor-general? Well, there is that and a co-
lonial war in space. In this section of the Real Deal, we aren't
going to explore some things that are not likely in humanity's second century in space (a British space empire for one),
but instead focus on some of the problems that the governor
brought up, which almost certainly will be factors as we reach
for the depths of our solar system.

Wireless Communication

We take wireless communication for granted here on good
ole earth. Well, I suppose I take wireless communication for
granted. If you are older than about 35, you might still remember being dependent on landlines. However, in space we don't
get 'landlines' or anything else, except wireless communication (if I need to explain why, then you and I might not make
good friends).

The critical limitation with wireless connectivity on earth is
the range a signal can go before a receiver unit cannot discern
anything useful because of atmospheric interference. This is
most commonly reflected in the 'bars' on the top of your cell
phone screen showing how good your signal is. In space there
is no atmosphere to break up signals. Therefore, for all intents
and purposes, they can keep going forever.[90]

While these signals can go on forever, they, like visible light, spread out as they travel. This means that even a very narrow beam laser fired from Mars to the Earth may end up being receivable over a large swath of the planet.[91] This isn't a problem, so long as we aren't worried about hiding our presence from hostile aliens. However, if you are a military that is attempting to hide your communications from a rival, it quickly becomes a critical vulnerability of wireless communication.

The militaries around the world already take this into consideration in their battle plans. They formulate backups that can't be intercepted. How will this be solved in space? Well, the easy answer is returning to the good old-fashioned courier. While it might take weeks to get a secret message between bases, getting it there in secret might be the difference between life and death. Alternatives include quantum communication,[92] although that remains decades in the future.

Transit Hubs

Have you seen the Expanse? If not, then again...we can't be friends. Since I assume there will be exactly three people who were not already my friends or mother who see this book – I'll explain.
The Expanse is a book[93] / TV series[94] about a future of space where there are three factions (Earth, Mars and the asteroid Belt) fighting over everything. I simplified it...but if you didn't care to watch/read it in the first place then you didn't want the long version. To help you out, for the rest of this story any background you will need is in italics.
Ok – that out of the way...those of you who have seen it (and the rest of you who now know everything you need to), what is the purpose of the station on Ceres[95] (*it is a station build into an asteroid*)? It is a transit hub. Sure, there are thousands of people living on it, but its core function is to refuel and supply operations happening in the asteroid belt. This isn't just a

nice story element for an otherwise awesome show. It is an absolute necessity for humans travelling away from the Earth's orbit.[96]

Our spacecraft, at least for the next 40-75 years, will remain dependent on stockpiles of fuel, food, water and other consumables to carry passengers beyond Mars. It isn't so much that a ship couldn't carry enough supplies for a multi-year trip, but the cost in space to do so will be seen as prohibitive.[97]

I'll paint the picture of what it would look like. Imagine that you are taking a road trip with seven of your friends in a mini-van. You have been looking forward to this trip for weeks, and these are some of your best friends. The first day is great, and you all are having a great time. However, instead of being able to get out and stretch, or have fun at the end of the day, you have to stay in the van. Instead of a two-day road trip to the beach, this is a six-month trip. Are you starting to understand more why it might not be a great idea to cram every inch of living space with food and water?

We will build these way stations, if for no other reason, to provide the fleshy little meat sacks that ride in our ships a chance to get off and waddle around. We will do it for the same reason we have movie theaters, bowling alleys and tennis courts here on earth. People need spaces to relax, blow off steam and socialize.[98] That is most certainly going to be in even greater demand up in the infinite black of space.

WHAT WILL WE DO?

Could you print that for me?

Remember the first time you found out that you could 3D print something that you previously thought had to be made by hand? For me, it was finding out you could 3D print models of ships. It was an emotional moment, because as a child I spent hundreds of hours on that. Oh well, it taught me patience. What if we could print larger ships? Let's check out how that might work, and the companies that are making it a reality.

The Fun Part

June 1, 2034
Space Solutions Lagrange One Shipyard 326,054 kilometers above Earth

Hey, hand me that wrench, will you? No, not that one. What exactly do they teach in trade schools these days? Whatever, I'll just teach you. It will be faster than that lousy six-month program you went through, and much better.

First off, this whole 3D printing system is built around this nozzle. If it's clogged, nothing's gonna work. That is where any troubleshooting should start. In our case, everything looks clean, although this nozzle has seen better days. I assume some jerk on the graveyard shift has been hammering his head against this. We print space ships up here. And somehow, I still get sent the people who can't graduate high school...

Think of this whole system as just a giant cake frosting tool. We checked the nozzle here, where the alloys clog all the time. Next are the lines feeding the liquid alloy to the nozzle. I've been here four years, and I've never seen 'em clog, unless the main nozzle is gunked up. You might as well just skip checking them. If something is wrong with the feeder lines, they will just be spraying molten metal in your face. It's kinda hard to miss.

From there, we have the heating elements and storage tanks. The tanks are like the feeder lines. If something is wrong, you have metal in your face. If that hasn't happened, they are fine. However, the heaters go out all the time. I'm convinced that corporate sends us used heating elements, rather than new ones. If I were you, just replace them once a month at a minimum. That allows you to stay ahead of problems and avoid this down time. You could replace the nozzle around then too, but that is expensive. Watchout, because we are audited on how often we replace that.

Let's just replace the heater and see if that clears things up.

——Ten minutes later——

Hey screw up, next time you call me about the 3D printer not running, try checking to make sure the storage tanks actually have stuff in there. I'm done here.

The Real Deal
3D Printing Spaceships

This month Relativity Space[99] announced its first contract to deliver satellites into low earth orbit.[100] Why is this being talked about here? That is a fair question because it seems like every other day a contract is being signed by some new rocket company. This is different because Relativity Space prints its rockets.

Relativity's Terran 1 rocket, which features a flexible architecture, was fabricated using Relativity's patented technology platform on its giant Stargate 3D printer, which features 18-foot-tall robotic arms that use lasers to melt metal wire and can help lower the part count of a typical rocket from 100,000 to just 1,000.[101] In addition, Relativity says they can print the rocket in less than 60 days.[102] This is nothing short of revolutionary, as it has historically taken months if not years to make a single rocket. Even now, it takes 62 days for SpaceX to refurbish its already used rockets.[103]

*Stargate 3D Printer…the humans next to it
are real and not pretend dolls.*

Zero-G 3D Printing

There is a difference between 3D printing rockets on the ground and doing it in space. However, Relativity now has its sights on 3D printing rockets elsewhere. Once it masters its automation process here on Earth, the company hopes to shrink its printers and ship them to Mars via rockets, to see if they can create vehicles capable of launching from the Red Planet using raw metallic materials.[104]

It is a short step from the designs and techniques Relativity is pioneering, to a company like Made in Space[105] adapting them to their zero-G 3D printers. Made In Space has created the world's first zero-G 3D printer, and in partnership with NASA has been testing facets of it since at least 2015.[106]

One of its key projects is the Archinaut,[107] "a technology platform that enables autonomous manufacture and assembly of spacecraft systems on orbit". [108] This would enable a wide range of in-space manufacturing and assembly capabilities, fundamentally changing how spacecraft are designed and built. This capability is also a virtual requirement in order to support key technologies, like space-based solar power[109] or space elevators.[110]

You can't be serious?

I can hear some of you claiming that my 15-year timeline is a little aggressive for this. However, consider that the cost of getting to space has dropped 100-fold over the same time frame,[111] and there are realistic projects underway to begin mining resources on the moon over the next 10 years.[112] Therefore, a five-year lag between beginning resource extraction in orbit, and the capitalization of this by commercial companies to build factories up there, isn't unrealistic at all.

This is especially true with 85% of the tech already being used here.

A New Environment

Today, we are going to be looking at Josh as he leads his rag-tag protest movement on the Moon. You may think this a silly topic, but we are going to have to wrestle with how we deal with these new environments we are entering. There are no easy answers.

The Fun Part

September 18, 2049 Tranquility Crater, Luna

What the hell were they thinking? Svetla thought, as she looked out the window of Tranquillitatis Core (TC). "How the hell did those protesters manage to get here…and why signs…why do protesters always need signs?" she muttered to no one in particular. The protesters Svetla saw outside the primary European habitat on Luna appeared to be a…well… group of environmentalists would be one way of describing them. They were dressed in what could only be called 'second hand' clothes carrying signs calling for an end to development on the moon and returning Luna to its pre-settlement pristine condition.

Josh was out there with those protesters. Well, that wasn't technically accurate. He was out there, leading those protesters. Josh had a long history of political agitation back on earth, and after quitting his federal government job back in 2023, he had done little other than drift from one protest movement to another.

It wasn't that he was opposed to mining on the moon. Ironically enough, he was one of the protesters who probably most supported mining operations off Earth. But not like this. Definitely, not like this. Nearly two decades of strip mining across vast swaths of both Lunar poles had sparked Josh's current protest movement. You could even see some of the damage from Earth with little more than a pair of binoculars.

Tellus' Legacy, that was what Josh had dubbed his ad hoc group of friends who had helped him organize protests at some of the company headquarters of the major mining companies. They had just wanted the companies to use some of the environmentally sound mining practices that had been developed at the end of the 1900s, and early 2000s.

Serendipity proved to be an ally, when in early 2049 Josh ran into an old friend, Greg, who had made it big in the orbital whiskey craze of the 2030s. Greg offered to bankroll the group's trip to the Moon to stage the publicity stunt that was happening now.

Maybe, thought Josh, that two million dollars won't be badly spent, if we are able to actually change something.

We might just be another story for a niche reporter to talk about. Well, we can hope can't we?

The Real Deal
Space Protestors

It is a long-held maxim in political science that protesters are a natural outcropping of humans grouping together. Even in large families, we have that one uncle, cousin, or sister who just 'has' to fight back against what everyone suggests. Thinking that our journey into space will change that, is to deny our nature as fancy monkeys.

Therefore, if humans are going to keep being human, what can we expect to be some of the new frontiers in political agitation? Well, space environmentalism isn't going anywhere. In fact, it has been going on for years.[113]

In 2013, the US House of Representatives proposed making the Apollo lunar landing sites a U.S. National Park, and once it was a Park it would be submitted as a UNESCO as a world heritage site[114] (because if there is one thing American's know it is that all our National Parks are international treasures). While technically this would have violated multiple international treaties (if it had been passed into law), the attempt was symbolic.[115]

What are we to do with the historic and pristine new worlds we are visiting? Do they exist only to support humanity? Do we pillage them as we have Earth, in the pursuit of making our birthplace more livable? By contrast, do we keep them in pristine condition, depriving the billions of people living in poverty of the development inevitable that trillions of dollars of resources would bring?

Getting to the Moon

These aren't easy questions and shouldn't be treated as such. But what is easy? Getting to the moon by 2049.

In 2008, it cost over $10,000 to put a pound in orbit. That year NASA set the goal of 'hundreds of dollars' to move a pound into low earth orbit by 2033, and 'tens of dollars' for that same pound in 2048.[116] Right now, barely a decade after that goal, SpaceX can put a pound into orbit for about $1,500. There are multiple launch providers who are coming into the market over the coming years, including Relativity who will reduce the cost even further.[117]

It is completely conceivable that NASA's goal of hundreds of dollars per lb into low earth orbit, will end up being the cost to get to lunar orbit by 2033. When that happens, it is just a matter of time before 175lb people like me start making regular trips there, even for things like a couple-day protest.

(At Least Some) Dogs Go To Heaven

So – you have decided to take the plunge and finally go on that space vacation your kids have been nagging you to go on. Who is going to watch your six cats? And what about the fish? Well, maybe just bring them along. Why not? Let's see whether there are reasons not to.

The Fun Part

December 23, 2034
Waldorf Astoria LEO, 373 miles above Canada

Katie was not having this. She had very clearly specified in her reservation that she was bringing a dog. It wasn't a gerbil and a fish. How in the name of all that was holy did the hotel mess that up?

Outwardly calm, she replied (to the front desk attendant), "I understand that you don't have any small mammal accommodations ready, but they are here, and there is literally not a choice but for me to stay here."

"But ma'am," the clerk replied "It isn't just that we don't have accommodations ready, but we don't have accommodations at all. We are setup to handle human needs and have the ability to adjust for fish and birds, but not dogs."

"I get that, but let's pretend for a minute that I am standing in front of you with two dogs, and the return capsule doesn't leave for six days. Can we pretend that for a second? I realize that is silly, of course I don't have dogs here…oh wait I do."

"Yes, ma'am bu…"

"Let me stop you there. You have two options. Authorize me to return home now and expose your hotel to the lawsuit I will file when I return or show me to my room and we find something that will work for Jelly. OK? Do you think those are reasonable options? Because unless you are prepared to eject my dog into space, there isn't another choice."

"Of course, ma'am, I will get you right to your room. I'll have maintenance meet you there to determine what can be made for you."

"Thank you. Please let your manager know that this is the last time I'll be at a Hilton. I know for a fact that the Marriott up-orbit from here is ready for dogs."

The Real Deal

The bottom line for the Real Deal, is that if you treasure your pet above your children then you might want to wait until more research is done before bringing them on your next astro-vacation. It doesn't matter how pet friendly your space hotel is. However, if you like to live life on the edge, you need your emotional support peacock, or you aren't the spouse attached to the cats, then just make sure you buckle them in for the flight. Because despite some conflicting research, it does appear that space is largely safe (or as safe as anywhere humans go) for animals of all types[118] (although no research has yet been done on the effects of space on elephants or tigers).

Does Zero-G Hurt Animals?

Most animals seem to fall into one of three categories when responding to zero gravity. Some freeze, seemingly hoping that the effect will pass on its own, or simply accepting the futility of struggle. Others do the opposite, flailing madly without stopping. The last and smallest portion, set about actually trying to figure out how to swim efficiently through the air.

Geckos and other reptiles have been observed to go into a sort of "sky-diving" pose while in space. This is mostly observed in tree-dwelling species, and it makes sense that zero g might trigger that kind of response in creatures that live high above the ground.

Fish are used to swimming with lowered gravity. Therefore, they take particularly well to zero-g. A fish's buoyancy can simulate a lot of the effects of weightlessness, and their bodies are built for swimming. Moving through the air while weightless has a lot in common with swimming through water.[119]

In case any of you just read that paragraph and thought "oh no why are the fish swimming in the air....they won't be able to breathe"...please go back and re-read it...the fish swim in water...we swim in the air while in zero-G....please try and pay more attention ok?

Great – so I can bring my pet...but how do I take care of it?

Taking animals into space requires special considerations. If you were to take a gerbil or other small animal up, what would you need? Traditional aquarium-style cages don't provide enough traction for them to walk around. Historically, space mice have wire mesh cages so their toes can grip a rougher surface.[120] Wood chips couldn't be used for bedding. They wouldn't stay in place. Gravity-feed water bottles won't work.[121]

Therefore, pick up a pressurized water container instead. Don't even think of using bowls of dry food. Switch your little friend to compressed food bars.[122] As to how to clean the cages, you will need a special waste containment system similar to what NASA designed for its orbital experimental animals to keep everything in its place.

But do they like it up there?

Fish and tadpoles swim in loops, rather than straight lines. This is because there is no up or down to orient them. However, they generally seem to be fine.[123] Baby mammals have a hard time because they normally huddle for warmth, and in space it's hard to huddle when bodies drift and float. It's also difficult for babies to nurse, when they can't locate their mother's nipple. Otherwise, most animals (that have actually been sent to space) seem to adapt pretty well, and fairly rapidly.

What happens if I bring Fluffy Back?

This is an interesting question, because when humans return to earth, they get seriously affected with what I have dubbed Gaia's Revenge. We can, therefore, expect similar results with at least some animals. However, the exact responses seem to depend on the type of animal and whether it was born in space or not.

The smaller the animal, and the shorter time they are in space, the faster they recover when returning to Earth. Some animals born in space seem to have problems determining which way is up when they return to gravity; while others don't seem to have any such issues.[124]

Walking the Orbital Runway

Before we go further, let's just agree that we are just monkeys who like putting fancy things on our bodies to make other monkeys think we are cool.[125] I'm not saying that as a negative. I have over 20 three-piece suits, and an entire bedroom devoted to my clothes. However, we are just the version of monkeys who figured out that we should use our thumbs to button clothes rather than just pick our noses.

We also shouldn't be so idealistic that we think venturing out into the cosmos won't involve us taking our sense of fashion (or lack thereof) right along with us. That would be like assuming that one of the first things set up on any new habitat won't be a distillery.

Space may be 'hard' now...but even the Wild West had fashion shows, and plenty of booze. We are going to bring all our desires, vices and luxuries with us to the stars. If we can't bring it, we will make it there.

With that understood, let's see what Evelyn is up to.

The Fun Part

May 2, 2034
International Space Station, 361km above Earth

Damn, I do look good. Evelyn thought, as she stared into the mirror in the rather cramped dressing room she and the 13 other models were using on the refurbished, and now commercialized International Space Station. Although, seriously, who is going to wear anything like this? This is like a cross between a deep-sea diver outfit, and something that classic singer...what was her name...oh right, Lady Gaga,[126] would wear.

It doesn't matter to me what this looks like. I am just here to make it look amazing and figure out how I am going to 'walk' this runway. Thank god, we have a rehearsal today. I sure hope I don't actually have to try and maneuver my way around out there.

This micro gravity is not being friendly to my face. I seriously look fat and skinny at the same time. This is weird. Maybe I could have had the half bagel for breakfast and just claimed it was the gravity here messing with how I looked, oh well. Maybe if they ever have another one of these shows, I will.

I have no idea how Virginie Viard convinced the penny pinchers at Chanel to stage this show up here. I mean we do have a pretty fucking awesome backdrop, so that has to count for something. I don't think even Karl Lagerfeld[127] could have upstaged Virginie after she has us on that tether, 'walking' with the curve of the Earth as our backdrop.

Wait a minute. I just realized something. I'll be wearing a helmet out there. Why did I need to make sure I did my makeup first? This is really strange…

The Real Deal

Is it really so implausible that one of the major fashion houses, who already spend in excess of $10 million to stage their annual fashion shows,[128] won't decide that it would be worth it to shoot some of their models into space to make a splash? Those fashion houses may be the first people to actually look at making space suits look 'better' than they do now.[129] It is only a matter of time before the number of people living in space will create a demand for distinct styles of space suits.

This picture is a design that NASA recently posted to their Twitter account.[130] Yes, it does look like a weird crop top. …However, NASA is actually looking to revamp their current suits. This is partly because the suits costs hundreds of millions of dollars. It is also because, as we saw earlier, they don't actually have them small enough to fit women.[131] It is also, in part, because the current suits are really hard to work in, much less walk in. If astronauts are going to be waddling around the moon, it would be nice for them to actually be able to do something useful.[132]

Drill Baby Drill

It has been quite a while since the Wild West days, where everyone was their own law and where enforcement was at the point of a gun. Asteroid mining is one place where that life could come back, short of some sort of structure to enforce claims[133] and the law. Let's look at that.

The Fun Part

Darn those claim jumpers; I should have killed them when I first saw them on my rock!

Calm down, Mr. Siemans. As your attorney, I should warn you that attorney-client privilege doesn't extend to criminal behavior. I am quite glad you didn't kill anyone.

But Jeff, this is the second rock I have lost to the same crew!

Slow down and explain what happened this time.

OK, I was just coming back from my second haul of ice to the Lagrange fuel depot there near the Gateway,[134] when I found another ship on the landing pad.

What ship?
It was a Water4You's ship. It was their latest made-in-orbit one. It looks really nice, but not when it is on MY mine!

How did you know it was Water4You's ship?

Because it was the same one that took my first mine six months ago.

Ah, where was that?

Between here and Mars

And you registered it?

No, I didn't have the money back then.

But did you register this second one?

Yes, since it was so much bigger, it meant that I would get more scrutiny.

O.K., go back to when you saw them on your asteroid. What did you do?

I went down and was immediately stopped by some armed thug, who told me to get off his company's property. I spent about an hour arguing with him. However, when my air started to run low, I figured I should head back to my ship.

O.K. What happened next?

When I got back, I had an alert on my navigation system that said I was in a restricted area. If I didn't leave immediately, I would be considered hostile. I could be targeted by any law enforcement vessel.

What did you do?

Well, I left. I didn't want any trouble with the law, especially because my sister has had a few run ins with the Chinese contingent up here. They aren't terribly friendly.

How did you find me?

My dad has used your firm for all his companies, so I came right to your office here at GEO1.

Ah, that explains why your name is familiar.

Great, so you have heard of my dad. Can we now find a way to deal with these guys? This is the second time I have lost a mine to them!

I understand Mr. Siemans, but now we have a case. Your first mine was technically unlicensed with the orbital guard,[135] meaning this is effectively the first 'real' mine that you lost. Since we filed a claim on this asteroid, we can now take-

Water4You to court. And not only will you get your asteroid back, but also damages which, unless I have misread W4Y's financials, will essentially compensate you for your first mine.

Are you serious Jeff? Do you really think we can get anything from these guys? I mean that crew out there seemed like a bunch of thugs. You are saying they are part of a legit company that we can sue?

Well, the company is legit yes. That crew...not sure. I am sure the company will try to claim that the crew was 'rogue'. However, since it is the same crew both times, and since they were flying W4Y's newest ship, I think we have a good case that they were sanctioned.

That is way better news than I thought I was going to get from you. I just came to you to figure out if you knew anyone who could 'take care' of those guys.

Space Mining Probably Won't Look This Cool (Image from the movie Armageddon (1998) - Touchstone Pictures)

The Real Deal

It has been a while since claim jumpers have been a problem for miners anywhere. However, we are likely entering an era when claim jumping on mines is going to be both profitable and possible, until a robust law enforcement presence is established in space. However, before we get to that part, let's explore space mining in more detail.

Orbital Roughnecking

Over the past 20 years, scientists have identified over 11,000 near-Earth asteroids (in addition to the near million we know of in the Main Belt).[136] A handful of companies are exploring options to begin prospecting on some of these asteroids in the coming years. Most of these plans involve initially mining ice to be converted into fuel (because refueling in space is much easier than bringing it all up from the ground).

After refueling infrastructure is more established, mining will likely start to expand to rare metals, and other minerals for use in on-orbit manufacturing.[137]

Despite the high number of near-Earth asteroids, researchers say the majority of these are not suitable for mining with current (or near current) tech,[138] because of the following five factors:[139]

Speed. One of the first considerations when looking at an asteroid to mine, is how fast it is going. The faster it is moving, the harder it will be for a spacecraft to match velocity and land. It is the difference between jumping on a train slowly pulling out of the station, or trying to jump onto a moving bullet train. James Bond may be able to do both, but that doesn't mean they are the same difficulty.

Spin. For many of the same reasons as speed, the spin of an asteroid can prevent a landing. However, aside from trying to land on a top, the faster an asteroid is spinning, the more likely it is to actually disintegrate, making it a less desirable target for mining. There is no point in risking equipment if it is going to be lost after the drill breaks the entire rock up.

Size. Yes, I am saying size matters. At least here. Anywhere else, you are on your own. In this case, it is because anything smaller than about 3-400 meters across, just won't have enough water or ore to make a trip worth it.

Structure. Asteroids are classified by their structural composition. C-type asteroids are about 20% frozen water (and considered ideal initial targets for space mining operations), M-type are largely metallic and S-type are stony.

Orbit. At the end of the day, the closer an asteroid is, the easier it is to get to. If you have the four S's aligned, but the asteroid is far away, then it is probably not a good prospect for your mining operation.

Great, but seriously…mining in space?

Interest in space mining has increased recently, as Musk, Bezos and others have publicly pushed forward in space. However, their efforts to reduce the cost of space travel is doomed without finding materials, such as water and minerals, that do not have to be rocketed up from Earth.

Goldman Sachs wrote a research note in 2017[140] stating:
> "Space mining could be more realistic than perceived. Water and platinum group metals that are abundant on asteroids are highly disruptive from a technological and economic standpoint. Water is easily converted into rocket fuel and can even be used unaltered as a propellant. Ultimately being able to stockpile the fuel in LEO [low earth orbit] would be a game changer for how we access space. And platinum is platinum. According to a 2012 Reuters interview with Planetary Resources, a single asteroid the size of a football field could contain $25bn- $50bn worth of platinum."

That same report suggested that within the next 5-10 years, we should expect to see an exponential rise in orbital mining interest from new and established companies.

Goldman was not the only large investment entity which has been taking notice of the potential for orbital mining. Since the middle of the 2010s, the United Arab Emirates and Saudi Arabia have been investing in ground-based infrastructure to launch satellites.[141]

But Are Claim Jumpers A Problem?

As for ownership, the 1967 United Nations Outer Space Treaty says that no nation can claim ownership of the Moon.[142] However, that doesn't necessarily prevent private companies from claiming portions of the Moon as their own commercial property.[143] There still isn't legal framework for claiming property elsewhere in space.[144]

Since regulation and laws usually take time to catch up to technology and the 'real' world, we can probably expect at least a handful of asteroids to already be getting mined before any regulations come out governing that. Realistically, until there is a problem (like one company stealing a mine from another), there isn't likely to be much impetus for lawmakers or regulators to actually do anything.

The moon is likely to get more attention in the near term, as companies look to establish bases there.[145] However, with many of the early companies heading for places far away from each other, there is the possibility that they will avoid coming into conflict for a while.

Nonetheless, ownership in space[146] is an ethical conundrum that space law experts and ethicists have not yet solved.[147] However, it seems like we might need to start thinking of a solution soon.

WHAT ABOUT SIN AND DEBAUCHERY?

If you are more than a year out of your college days, the thought of being upside down while drinking your body-weight in beer isn't cool anymore. However, if you are even a little into socializing, you understand that alcohol is an important tool for our monkey minds as we seek the company of others. Do astronauts drink now? Does alcohol work differently in space? Let's see what Tony is up to and learn more.

The Fun Part

October 31, 2028
Lagrange 1, ~200,000 miles from Earth

"Darn," thought Tony, "where am I?"

Oh, right. Lagrange One. Darn, my head hurts. What happened last night?

As his mind struggled to put together a picture of what happened the night before, Tony noticed what could only be a semblance of vomit on the air vent above him.

"I am never going to hear the end of this," he thought as he began to remember the celebration his team had after their first successful rendezvous with one of the company's lunar fuel shuttles.

The crew had been on the station for about six weeks, monitoring the automated fuel processing station that had been set up on the moon earlier in the year.

Tony's crew was the first to actually receive anything from there, after months of delays. However, now that the system was up and running and converting ice to rocket fuel and water, there would be a shuttle arriving roughly once every 10 days.

"Man, I never thought I would be a hundred thousand miles out and getting wasted" Tony mused. "Then again...why did I think that Stella *wouldn't* bring booze for us?"

I just hope the rest of the team got similarly wasted. There is nothing like being the only one. There isn't enough happening up here for them to forget.

The Real Deal

The idea of drinking alcohol in space has been around at least since man landed on the moon. That's when Buzz Aldrin took communion and drank a glass of wine.[148] NASA doesn't really want you to know about that, more so because of a separation of church and state, than because there's anything wrong with an astronaut consuming a little vino.[149]

*Neil Armstrong while Buzz was taking communion
(Image originally a Calsberg beer ad)*

However, Aldrin did drink alcohol in space. This proves it can be done. He wasn't the only one to do so. It just isn't allowed, at least not by NASA.

History of Booze in Space

In the 1970s, in order to combat what was seen as the terrible food being served to astronauts, for a very brief time, NASA allowed sherry[150] to be consumed during Skylab missions. The reason for selecting sherry, according to "The Astronaut's Cookbook," [151] was that according to "several professors at the University of California at Davis, it was decided that a sherry would work best because any wine flown would have to be repackaged. Sherry is a very stable product, having been heated during the processing. Thus, it would be the least likely to undergo changes if it were to be repackaged."

However, it never made it. During a public lecture, Skylab 4 commander Gerry Carr happened to mention the inclusion of sherry in the meal packages for the next mission. The public was so enraged about astronauts consuming alcohol, that the idea was scrapped before it ever got off the ground.[152] Literally. Since then, NASA has had a very strict ban on alcohol in space.

The Russians, on the other hand, have historically had a looser policy, with doctors recommending the cosmonauts on the Mir space station actually drink Cognac to keep their immune systems healthy.[153] There is some scientific evidence to support this claim. A 2011 paper[154] found that resveratrol "could be envisaged as a nutritional countermeasure for spaceflight." It is good for you.

Despite the Russians' encouragement of alcohol consumption in space, the International Space Station is a dry facility.[155] Well, it is mostly dry. The astronauts there did actually brew beer as part of an experiment in the early 2000s.[156] And there have been other alcohol related experiments, including testing how whiskey ages in space.[157]

What will we drink in space?

Beer and champagne are poorly suited to space parties. Why? Because of the gas that makes the little bubbles in them. Without gravity to draw liquids to the bottoms of their stomachs, leaving gases at the top, astronauts tend to produce wet burps.[158] Which if you have never heard or experienced, are neither fun nor attractive.

This obstacle isn't stopping others from trying to troubleshoot beer in space. One startup called Vostok is trying to develop the world's first space beer in order to tap into the space tourism market early.[159] It's currently working out a formula of a low-carbonated beer that will reduce the chances of contending with wet burps.[160]

You might think that beer in space sounds great. However, in all likelihood, private companies that want to offer space tourists a little something for the ride, will likely just stick to wine and liquor and avoid the bubbly mess.

"That's one of the reasons why we don't have carbonated beverages on the space menu," said NASA spokesperson William Jeffs, who clearly didn't want to talk about why the space menu didn't include fun drinks.

From left are Hibiki 12-year-old, Yamazaki 18 and 12-year-old Japanese whiskeys, three of the types of whiskey Suntory sent to space (AP Photo/Eric Risberg)

OK, so if beer and champagne are out, and vodka isn't your thing, then maybe just stick with whiskey. Suntory, a Japanese distillery, has sent a number of different types of its whiskey to space, in order to study the effects of micro-gravity on the flavor.[161] No word yet on the results, but with Suntory winning a number of awards for having the best whiskey on Earth, they may be looking to seize the best whiskey in space award early.

Won't getting drunk in space suck?

There is a widely held belief that getting sloshed at higher altitudes makes you feel woozier faster. It would, therefore, seem logical to assume that drinking alcohol while in orbit could have even more bizarre effects on humans. However, this notion may not actually be true.

In fact, there is evidence that debunks this myth dating back to the 1980s. In 1985, the U.S. Federal Aviation Administration conducted a study[162] that monitored whether alcohol consumed at simulated altitudes affected performance of complex tasks and breathalyzer readings.

In the study, 17 men were asked to down some vodka both at ground level and in a chamber that simulated an altitude of 12,500 ft. They were then asked to complete tasks including mental math, tracking lights on an oscilloscope with a joystick and a variety of other tests. The researchers found "there was no interactive effect of alcohol and altitude on either breathalyzer readings or performance scores."

The idea of altitude affecting drunkenness probably comes from people who claim to have gotten drunk on a plane, in a way that's faster than normal. This is probably the "think-drink" effect,[163] which has been studied extensively over the years. It suggests that people are going to act more drunk if they think they are drunk.[164] They don't even have to actually consume alcohol.

That might be the good news part of this whole section. You still can drink in space and it probably won't mess you up too much, assuming you don't work for NASA, those prudes.

Shaken Or Stirred?

Have you ever had a cocktail? I know that it is a silly question. But have you ever really paid attention to how they are made? Do you think that making them would work, if there was no gravity? Hint: the answer is no. Then what are we supposed to do up in space if we: (a) aren't sober or (b) don't like just straight liquor? Well, let's see a promising solution.

The Fun Part

The InterContinental Geo
Geostationary Orbit, 28,000 miles above NYC

"This was never a problem on the moon, which is why we never came up with a new way of doing this," Jessica explained patiently. "When I was a bartender there it may not have been easy to mix drinks, but it sure was possible."

Ken, Jessica's boss, was eyeing the new device Jessica had brought over from the science facility. "Just what do you plan on doing with that?" he asked.
"Well, for starters, actually make a drink that people on this floating hotel will enjoy. We already can't serve beer, because of how it messes up people's stomachs. While wine and liquor are all well and good for you old people, some of us actually want more than one flavor in our drinks."

"Yes, I get all that Jess. However, this looks like a $20,000 device. How can we justify buying one of these, just so we can sell a handful of more types of drinks?"

"First off, Ken, it isn't a handful of more types. It is hundreds or thousands more. Second, even 30 years ago, this was an $800 device[165] back on Earth. I am pretty confident we can get a non-science grade version of this for under $1,000."

"Oh well, $1,000 is an easy sell. Next time, lead with how

cheap it is going to be." "Right, next time I come up with a game changing way of mixing drinks in zero-g, I'll do that."

The Real Deal

There are two problems with cocktails in space or really just drinking in space (other than the current NASA ban[166]). The first is mixing drinks, and the second is drinking the drinks. While they are both physics problems, the first requires a more technologically sophisticated solution than the second. Let's start with the mixing.

Martini...Spun Not Stirred

How do you do it, when there isn't anything that pulls the liquid down? You might not think that is an issue, but look at what is happening when you stir your coffee in the morning. There is a downspout that forms, pulling liquid down. That won't happen in space, so now what?

Enter the centrifuge, a device that humans have been using for decades to make weapons of mass destruction. You probably used it in a college chemistry class to extract solutions from water.

As early as the mid-2010s, bartenders were experimenting with using centrifuges for making cocktails as a novelty.[167] However, in space, these devices are likely to be our only way to mix drinks, until we can create artificial gravity.

Wait, but how does a centrifuge work?

Yes, this is a centrifuge that you can buy on Amazon[168]

Oh right, some of you were able to get by on your good looks or sports skills and never had to learn about science. A centrifuge is a device that spins liquid samples at high speeds. It creates a strong centripetal (fancy word for spinning) force, causing the denser materials to travel towards the bottom of the centrifuge tube. In space, a centrifuge can be used to recreate the effect of gravity, by pushing things against the wall of the centrifuge tube.[169]

One nice thing about current centrifuges, is that they have individual compartments (like in the picture here), so you could theoretically have multiple drinks mixing at once.

Cool But What About The Booze?

Let me get there, gosh. Creating the artificial gravity inside the centrifuge is half the battle, since it at least creates the situation where the liquid can be mixed. Our little space cocktail machine would also need an agitator or mixer of some type to ensure that the liquid was simultaneously being gravitized (yes I made up that word...if President George W. Bush could,[170] then so can I) and mixed.

Bottom's Up

The coolest martini glass you will see this year

If you take away gravity, a liquid becomes something different. It is sticky. Surface tension keeps it clumped together into blobs.[171] If you shatter that blob, dozens of little blobs scatter everywhere. It then sticks to your clothes, skin and everywhere. It is hard to clean up. The current space technology solution for managing liquids is simple: keep it in a bag.[172] Squeeze bags are ubiquitous. You use them for camping and travel. They are practical. They are also ugly,[173] which apparently outweighs their utility.

A Kickstarter back in 2015 devised a solution (image above) to this problem, by breaking up the inside of a martini glass with grooves to prevent the liquid from forming into blobs, and instead be guided to your mouth. I think the important thing to take away here is that (1) drinking in space will happen and (2) we have our 'top' people working on solutions to any problems with drinking in space...[174]

Do We Have Enough Space to Get It On?

Let's face it, the concept of space-sex is pretty hot. When humans eventually strike out to create colonies on other planets, we're probably going to need to procreate, even on the way there. How complicated will it be to put said bun in the oven, while we're on our way to another planet?[175]

The Fun Part

January 11, 2029
The Gateway, Lunar Orbit

Well, how was it?

What?

Don't what me. How was he, you know, in bed?

One, we weren't in bed....Two, it is none of your business.

I'm your sister. Of course, it is my business.

Fine, it was gross. Yeah, I don't really have another description for it.

Gross, what do you mean?

Well, it was a bit like having sex in the middle of a kid's pool. However, the pool is filled with sweat and it is 100 degrees outside. You can't actually do much screwing, because you are floating.

Hmm, I had imagined sex in space to be a little more fun. Maybe you are just bad at it. You were always the one who had a hard time with men.

Screw you, Kylie. I may not be as...social... as you. However, I do just fine with men. Well, so long as I have a literal leg to stand on.

Oh God, Jess. That joke was worse your attempt at making pancakes up here.

The Real Deal

Sex in space is a logistical nightmare, with problems ranging from floating fluids to shrinking manhood's, according to a NASA-funded scientist.[176]

John Millis, a physicist and astronomer, recently discussed the issues faced by consenting astronauts who wish to engage in naughty acts in micro-gravity. Millis, of Anderson University in Indiana, compared sex in space to having intercourse while skydiving.[177] It is difficult but not impossible.

"The issues surrounding the act all revolve around the free-fall, micro-gravity environment experienced by astronauts," he said. "Imagine engaging in sexual activity while skydiving, every push or thrust will propel you in opposite directions.

"Even the lightest touch can make it difficult to stay in contact, if both persons are not properly anchored. The astronauts would need to brace themselves against the space station and even each other. "A shared sleeping bag, or similar, would perhaps be the most useful."

Locking bodies is only the first problem

In micro-gravity, blood rises to your head,[178] instead of your genitals. This makes it harder for both men and women to get aroused. This low blood pressure below the waist also causes the tissue in a man's penis to shrivel.[179] This potentially impacts an astronaut's confidence, when it comes to liftoff.

Another issue is that male testosterone levels plummet in space. This means that rocket-riding adventurers suffer from a lack of sex drive.[180] In fact, it appears that the heart will shrink over time, [181] the longer astronauts are in orbit. That means there is less blood in our lower extremities to be used for fun.

Surely there isn't more bad news…

Sex in an environment lacking gravity would result in all fluids such as sweat, vaginal wetness and semen pooling and floating around the cabin.[182] This would make the high-jinx more than slightly uncomfortable. (Assuming you don't enjoy having sex in a pool of your own body fluids.)

Because of the micro-gravity environment, sweat and tears don't run down the astronaut's bodies like it does here on Earth.[183] Instead, it pools like small ponds of fluid near where it was secreted.

OK, so how about some solutions?

In his book, "Life in Space," [184] NASA technician Harry Stine claimed that sex had been simulated by the space agency. They found that a lucky "third person" was the best way to facilitate the friskiness. No, not like that. Think more anchor than participant.

Testing 2 suit in zero-g

The third person would serve to hold/push one of the participants against a wall, while that participant holds the other one.

Sci-fi author and inventor Vanna Bonta created an outfit known as the 2suit[185] to prevent space-sexual frustration. Bonta came up with the idea, while she and her husband were taking part in a zero-gravity flight and were unable to hug each other, because of the weightlessness.

The 2suit has large flaps which open at the groin and is covered in Velcro. This enables users to attach to each other and engage in sexual intercourse.

This...like alcohol...is one of the top problems of the space tourism industry is going to be grappling with over the next decade or so[186]...after all, who wants to travel somewhere were drinking and carousing isn't feasible?

The Highest Roller

What is more fun than going on a nice vacation and losing all your money? It is doing that in space! People have been going to Vegas for decades just for the thrill of pretending they can win. For centuries before that, people have been gambling away their wealth. Therefore, what better way to usher in the Space Age, than to ensure it includes a casino?

The Fun Part

April 3, 2053
MGM Lagrange, Approximately 326,054 km from Earth

Maybe I do have a problem. I have been up here what, three weeks now? When was the last time I called home? Oh God, have I called since I got here?

Hmm, but if I just do one more hand, that is all I need to get back on top. I know my luck is starting to turn.

Mr. Phillips, I've been asked by the front desk to inform you that your card has been declined for your room charge. Do you have another card we can use?

Uh, yes, try my American Express. Here it is.

Thank you, sir. Enjoy yourself. Do you need anything?

Something to eat would be great. It feels like a week since I last ate.

Very good, I'll have a hostess come by with a menu.

Thank you, what was your name again?

Alan, sir.

Oh right, thank you Alan.

It is my pleasure sir.

Damn it, where was I? Oh right, if I put $100 on 26, $1,000 on black, and $300 on odd, that should get me back on top.

Just one more spin of the wheel. Just one more is all I need.

The Real Deal

So, I'm going to assume you all know what a casino is. Therefore, I don't need to explain how a space version of it would work. Well, at least not the casino part. The games are a different matter.[187]

Poker, slot machines and blackjack will all probably look pretty similar to earth. While digital versions of both poker and blackjack are currently available, they aren't used in casinos. This is because we are silly monkeys, who trust being able to touch things. That will probably persist in space, although I'm not ruling out the first space casino being all digital.[188]

Other games, like roulette and craps rely on gravity to work. Before you say just use magnets, you should remember that people play those games, because they believe there isn't an outside force causing them to lose. Well, other than luck and God that is. However, they believe that the dice and table are level and not weighted or otherwise skewed in the house's favor. So, if you add magnets to the table or dice, then they can be manipulated by the casino without people seeing or knowing. That is unlikely to be accepted.

Magnets are probably not the solution, other than for keeping chips on the table (more on that in a second).

Craps

An easier and more palatable solution for craps, would be a small dice gun which shoots out the dice towards the table using compressed air. This would propel the dice, while also ensuring the dice continue to roll until they hit the table.

After the dice hit the table, however, you need something to keep them from bouncing right back off and out into the room. That's where table fans come in, blowing air down towards the table at speeds high enough to allow the dice to bounce, but not fly off. These fans would need to be connected to the dice gun to know when to turn on (no one wants a strong fan blowing their hand, while they try to get their chips in place).

What about the chips? Oh right, back to that. The chips are going to need to be magnetized to keep the fans from blowing them around. However, instead of the table being a permanent magnet, they would need two adjustable electromagnets in them to ensure at the end of each round the chips can be easily swept up, and new ones put in their place.

Roulette

Roulette is a harder game to adapt, because the centrifugal force that keeps the ball rolling around the edge, will also prevent the ball from bouncing down. Unlike on earth, where the design of the wheel allows the ball to spin around but ultimately settle down, a space version would need to use a similar application of forced air to gently push the ball down, as the wheel begins to slow. I know that it is not terribly exciting. However, the alternative is an all-digital version of the game.

WHAT WILL HAPPEN TO OUR BODIES?

Sleepless over Seattle

Yeah, sleeping in space doesn't look quite like that now. It may not for a while. However, what does it look like? And how do people sleep in space? Turns out, them sleeping looks goofy, and worse than that they don't actually sleep well up there. Let's check out what is happening with Kendra, and learn more.

The Fun Part

August 9, 2024
En-route to the Gateway, 213,000 miles from Earth

Kendra, how do you sleep so well? I mean, it seems like you can fall asleep anywhere in here.

What do you mean, Dan?

I feel like you are the only one of the four of us who has actually slept more than about an hour since we left the ground. I know I have been getting just little cat naps because I just can't deal with all this wrongness. Where do I put my hands? Where do I sleep when everywhere is the same as where I eat work and the rest? How do you do it?

I don't know what to tell you Dan. I just sleep. I go to my little cubby and sleep. I've slept in worse places.

Worse places! How can you say that? We have no gravity. We have about as much room as a prisoner. We don't have night. We don't have a bed. How could you sleep in worse places?

Oh well, not this different. That is for sure. However, I went to the Army's Survival, Evasion, Resistance and Escape School.[189] We had to sleep in confined spaces with loud noises at all hours, being battered about the head and face and fed just enough to stave off actual starvation. In my head, so long as I've got a full belly and am not getting my face punched right before I go to bed, then I've slept in worse places.

OK, fair. However, that was what, like a week tops. We have been up here for almost a week now. We will be for another couple weeks. How can you be so calm about this?

I don't know. I don't think of it as being calm. I just have my sleep mantra[190] that I have used since I was a kid. Regardless of where I am, it helps put me to sleep. Bringing this sleep mask didn't hurt either.

Sleep mantra?

It is like a meditation chant, or some other word or phrase that focuses your mind. You haven't ever tried yoga or meditation, have you?

No, I haven't.

You just repeat a word or phrase that you associate with a mental action you want to take. For me, it is just repeating 'sleep'. It is like my mental 'sleep' word. I repeat that in my head for about five minutes and I'm out.

It seems a little wonky. However, at this point, I'll try just about anything other than the drugs they sent up for us to use to sleep. That stuff makes me feel awful.

I am not going to tell you what to do or anything. Although as the commander, I guess I could..... I'm just kidding... I don't think I actually have any authority up here You should have seen your face. My point is that we need you to be ready when we get to the Gateway. Therefore, whether it is my mantra or the drugs. You need to get this sorted out soon.

The Real Deal

You might assume that it's entirely pleasant to sleep in space. You just float. However, astronaut Mark Kelly points out that this actually makes snoozing *more* difficult. Interviewed after his 340 day stay on the international Space Station, Kelly said "Sleeping is harder here in space than on a bed...because the sleep position here is the same position throughout the day. You don't ever get that sense of gratifying relaxation here that you do on Earth after a long day at work. Yes, there are humming noises on station that affect my sleep, so I wear ear plugs to [bed]."[191]

Tired of the Problem

It turns out that Kelly was not the first astronaut to complain about how hard it was to sleep in space. A 2014 study[192] found that 85 astronauts, across almost 4,300 nights, were chronically sleep deprived. Many of them resorted to sleeping pills to get and stay asleep.

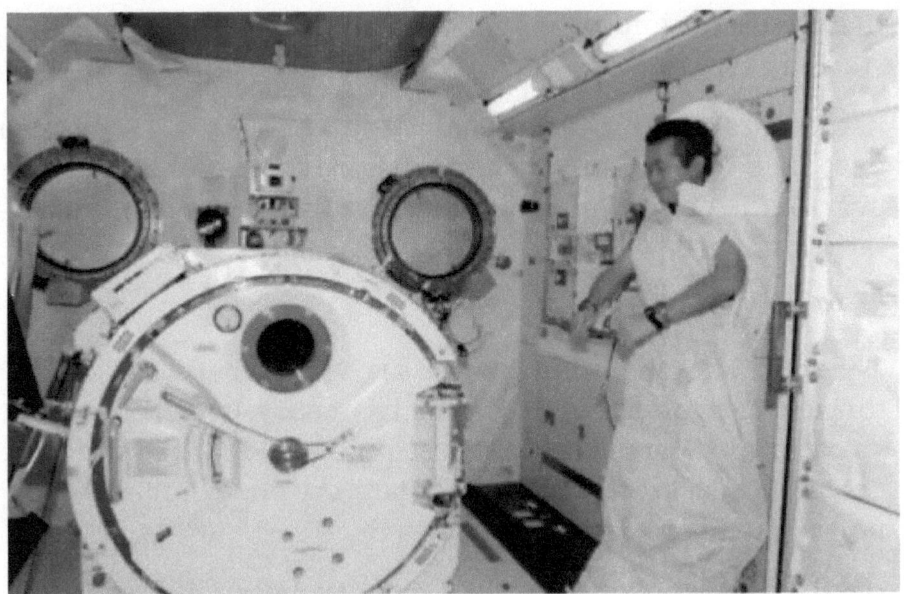

Not sure, that I would sleep so well like this either.[193]

This is not exactly an encouraging statistic for those of us who would like to travel to Mars or beyond. These astronauts are doing all of the things that would help us sleep better on Earth. They are exercising (two hours a day), doing meaningful, fulfilling work and generally abstaining from alcohol. Therefore, what is it that causes them to have such a hard time?

Researchers believe a large part of the problem is the unfamiliar environment, which causes a disruption of our circadian rhythm.[194] There is also the isolation, a sunrise every 90 minutes and a 'robust' ventilation system, which makes the ISS quite noisy.[195] It is a perfect storm of problems facing these sleep-deprived travelers.

Research around sleeping in space has been ongoing for decades. Some of the more recent studies have found that even just the isolation is enough to cause sleep disturbances in some.[196] Other studies have shown that astronauts have about a 24.2 hour internal 'day'.[197] While this doesn't seem like a huge deal, it means they are an hour off wherever they started every four days.

Astronauts currently undergo sleep therapy.[198] They also have rigid schedules to try and mitigate these problems. However, this isn't a sustainable solution, as we increase our population in space.

So What Will We Do?

The most effective solution is also the one that will take the longest. It involves making artificial gravity. It will enable the variation in work, play and rest postures. By that time, we should have a greater ability to vary the environment (light, noise…etc.) between compartments.

Until then, the solution will have to be a rigid schedule, combined with separating sleep and work compartments.[199] Astronauts already have all of the symptoms of some of the worst shift work sleep disorders seen on Earth. Therefore, applying the solutions found here will likely help up in orbit.

This includes creating a day/night cycle through lighting. It also involves being extremely vigilant about the consumption of stimulants and judiciously using herbal supplements, like melatonin. It will also help to integrate relaxation exercises, like Tai Chi or yoga to calm the body close to bedtime.[200]

Gaia's Revenge

There is a lot of talk about the dangers of space travel, such as the risk of radiation causing sterility, DNA damage or other maladies. However, much less is discussed about what happens when space travelers return to Earth after being away. We will check in on Tyler after his return from a six-month mining trip. We will see just what revenge humanity's birthplace is exacting on him.

The Fun Part

January 2032
Approximately 0.5 feet above San Francisco on a couch.

Six months was all. Why was he feeling like such shit? Just getting up from the couch seemed to ignite a fire in his knees and back. Laying there felt a bit like he was being crushed.

Tyler's combination of prior military experience, a dad who was a coal miner and a degree in electrical engineering made him a shoe-in for the first mining trip to the asteroid *1999 JU3*. Planetary Resources had been planning that first trip for five years, and man they paid well.

By any reasonable metric, that first trip was a fantastic success. The company sold the cargo for well in excess of $1.5 billion. Since anything above a billion meant his team and the company split the proceeds 50-50, he was sitting on an extra 45 million dollars. Not a bad half-year wage.

However, he couldn't enjoy that money laying here on the couch. What was wrong with his legs?
They looked like balloons. He had been cleared by the company's medical team three days ago after he got back, so what was going on? They told him that he 'might experience flu-like symptoms for a while', but this was clearly more than the flu. Although come to think of it, he was having a hard time

concentrating.

Maybe it would help if he just got up and got active. After all, just laying around wouldn't make him better. Tyler had been an active runner, and a closet Crossfitter before he went to 1999 JU3, but going on a run now seemed out of the question.

Eating didn't seem like a good idea either. He had already thrown up twice today. What about just sitting up and watching something? That would be a good start, right? As Tyler started to sit, he felt a horrible itching on his side. He pulled his shirt up to reveal his entire side covered in a bright red, splotchy rash.

OK, he thought. That's it. I'm calling 911. Something is really wrong here.

The Real Deal
The Returnee's Syndrome

The problems Tyler is experiencing are pretty normal for people coming back from extended sessions in micro/zero-G environments. This month, NASA published the results of their 'twin study' in the journal *Science*.[201] The study found humans are able to adapt to living in micro-gravity, despite the many hazards from increased radiation exposure and weightlessness.

Nevertheless, there are significant challenges for the human body, including hypoxia, disrupted circadian rhythm and abnormal gene activation (the full effects of this are not understood yet, but no our DNA doesn't change in space[202]). The researchers theorize that being further from Earth's protective magnetosphere, such as for a trip to an asteroid, would increase the damage to our bodies.[203]

While this study is important for understanding the impact of space flight on the human body, a particularly interesting thing happens to humans when they return groundside. [204] Their body rejects being home.

Astronaut Scott Kelly's (one of the twins in the study) immune system went on high alert upon returning to Earth (it did the same thing when he first went into orbit). His body acted as if it were under attack.
Mentally, Scott experienced at least temporary (roughly six months) cognitive decline. He likened his experience of trying to work to taking the SAT while having the flu. He may have known the answer, or how to do something, but accessing the information in his brain was problematic.

Scott also noted that adjusting to life in space was easier than re-adjusting to living back on Earth. He said he didn't feel like he got back to 'normal' for about 8 months. In the days and weeks after he got back, he had all of the symptoms that Tyler did above, and then some.

It isn't clear if some of these conditions were worse because of Kelly's age (he was over 50 when he spent his year in space), or if he has some genetic pre-disposition to Returnee's Syndrome. However, it sure looks like Gaia took her revenge on him for being gone so long.

Scott Kelly after his return from a year in space.
Yes, he was bald before he went up.

It may not be a fun thought to realize that returning to Earth after a trip to space will be a painful experience. This is especially true, considering how isolated space travelers will be.[205] People returning to society will likely want to get out and enjoy themselves. However, will Gaia's Revenge ensure that isn't possible? Maybe there is an easy medical fix for this condition. It may be the inevitable cost of a new environment.

If it isn't a preventable condition, it may be the spark which pushes people to form orbital societies faster than currently anticipated. I know that I would probably prefer to live with everyone I know and love in relative health up in space, rather than return and suffer for over six months. I guess we will see how prevalent this condition is, as more people venture forth from our home.

Habitat Fever

For most of us, being stuck inside for an extended period of time is a choice. Some prefer it, some don't. However, for virtually all of us, at some point we need to see the sky, experience other human, and socialize. Without those things, we feel the sensation of 'cabin fever'.[206]

These are not things we can count on experiencing once we venture out into space. For starters – the sky may just be the endless black of space. There will be more on that later. Now, let's see how Mike is doing.

The Fun Part

Mike liked to think of himself as a self-sufficient person. He had grown up in the slums of what had been Memphis. Well, he thought, it was still there. It just looked a little different, now that the Mississippi had expanded. Even before that, Mike had been really good at living off the land. His dad had taught him to hunt. His mom made him cook whatever he caught. The lack of entertainment growing up, meant he was supremely comfortable outdoors.

None of that had prepared him for what he found as he stepped off the Space X starliner, *Gateway to Mars*, a month ago. The small cluster of dome shaped houses seemed more reminiscent of a camping tent he used to use in the woods, rather than permanent homes. He saw what he assumed was an ice processing facility about 200 meters away from any of the other buildings, and a handful of greenhouse looking domes. Other than that – the entire colony seemed to have no more than 15 buildings for all 100 people.

It had only been a month, but things had changed so much. It seemed like yesterday that the thought of putting his feet on firm ground was going to solve all his problems. However,

Mike now realized that had just been his mind trying to cope with the reality of a six-month trip with 50 strangers.

It now seemed like he had no choice. It wasn't like he didn't like most of the people here. It was just that he had to do something. Without action, he was going to lose the last bit of what he wanted more than anything: autonomy.

Mike laughed, as he double-checked the drill. It was ironic wasn't it. He came here looking for autonomy. He found the combination of a militant sense of order and near total isolation overwhelming. He knew he wasn't the only one who thought so. However, why weren't the originals doing something about it?
Those first ten colonists had been here for almost two years. They had gone native and couldn't be trusted to actually run things, now that real people were here.

It wasn't so much that he had a problem killing them. Mike couldn't put into words how he felt. There was a vague feeling in the back of his mind that something was missing. Maybe it was just his old instincts reacting to him not having a real weapon. Either way, the time for action was now. The rest of the new crew was waiting for him. Once the originals were out of the way, things could finally get right again.

The Real Deal

Social order is more fragile than we think.[207] It depends on the tacit agreement[208] of the majority of participants. It is reinforced through social interaction.[209] A combination of hard living, limited resources and isolation may create the ideal setting for collapses of early versions of Martian society.

The Astrosociology Research Institute[210] is an organization dedicated to developing the nascent area of astrosociology. The institute was founded by Jim Pass, who received his PhD in sociology from the University of Southern California in 1991.

In 2003, Pass created a website focusing on how sociology can apply to space travel. The institute was founded in 2008 and lists a number of academics as part of its research team. Its *Journal of Astrosociology*[211] has already published two volumes.

Pass views cabin fever as an inevitable side effect of space travel, as we currently conceive of it. [212] The long travel times, small ships and austere habitats are almost guaranteed to wreak havoc on the human psyche. Pass advocates a more thoughtful approach to how we do this space thing.

We can't count on things 'just working out'. There may be similarities between space and ocean travel.[213] However, the critical difference is that at the end of the day sailors, colonists and travelers could socialize anywhere. Being limited by where oxygen and heat are present, is going to be a bigger problem than we may now realize.

"As the population [on Mars] gets larger, it also becomes more stranger oriented," Pass says. "In a small group, people know each other pretty well, but then as a group grows, then it becomes a situation where people start to form their own social groups and relationships, and they tend to isolate themselves into those kinds of structures."

For this reason, Pass is worried that some concepts for future Mars settlements don't contain enough socializing spaces. A central dome would enable greater interaction, perhaps with a park or other amenities.

Current Martian Housing Plans

In 2018, NASA and its partner, Bradley University of Peoria, Illinois, selected the top five teams to share a $100,000 prize in the latest stage of the agency's 3D-Printed Habitat Centennial Challenge competition. Winning teams successfully created digital representations of the physical and functional characteristics of a house on Mars, using specialized software tools. The teams earned prize money, based on scores assigned by a panel of subject matter experts from NASA, academia and industry.[214]

The actual designs of these habitats varied, as you can see in the photo gallery above, but all shared a similarity – a lack of communal space. This isn't a critique of the teams' submissions, but instead the idea that all humans will need in a living space is protection from the elements. We are intrinsically social animals. We have yet to live in an environment, where we couldn't just gather outside. As we move into space that is no longer an option. We will have to build social spaces into our environments.

A related issue with these hypothetical dwellings, is the heightened risk of isolation. Trips from Mars to Earth can take between three to six months, and even radio signals take around 20 minutes. It's not the same as spending months in the space station or on the moon, where communication can still pass back to earth quickly. Home is also a (relatively) short space flight away.

Martian colonists will not have these luxuries. This means their living situation will require much more careful planning. Putting groups of people into such an environment without a rigid structure, is going to create ripe conditions for chaos.

The early inhabitants could simply follow a command structure, similar to NASA or the military. However, over time, a growing population would require the inhabitants to divide labor, similar to villages on Earth.

This isn't meant to be a dire warning of inevitable doom. Instead, we should remember we solved the social order issue here on earth, and with planning, we can do it again, as we venture further from our birthplace.

Human in Name Only

It turns out that there is a surprising lack of information on the possible evolutionary trajectories of humanity once we leave earth. How could anthropologists, evolutionary biologists and others neglect this? Thankfully, one researcher has some theories. While they appear sound, they aren't great news for the unity of the human race.

The Fun Part

April 21, 2087
Korolev Crater, Mars

You can't be here, Alan. You know my parents have said I can't see you.

I know, Faye. However, I can't do this anymore. I tried going back to Earth, but the women there don't compare to you.

That's a tad trite, Alan. You know I appreciate the sentiment, but there is no need to go over the top like that.

OK fair. Faye, I am serious. I know that there are difficulties with us being together, but it is what I want. I think you do too.

Of course, I do, Alan. However, I also want to have a family. How can we do that? How are we supposed to have children? I'm third generation Martian, and you are Canadian, like real Canadian.

I'm sure that we can figure something out. Maybe in-vitro, or something else like that. I mean, it is almost 2090. Someone has to have a solution for this problem. We aren't the only ones like this.

Maybe not, Alan. However, you've seen how my family lives here. We are a tad on the...radical side...we don't even use tech that originated on Earth, unless it is derived from the first settlement. So I'm sure there is someone on Earth who has solved this, or maybe even someone on Luna, or one of stations at the Lagrange points. We don't have it here. There hasn't been a need. We have enough genetic stock, so no one has cared.

Don't do this to me, Faye. I really do think we are meant to be.

We might be, Alan. However, even things that are meant to be, aren't. It looks like we are one of those things.

The Real Deal

If that sounded a bit like Romeo and Juliet on Mars, then you may have paid too much attention in your literature class. Joking aside, there is a grain of truth to the idea that two families could forbid marriage (and procreating) between themselves, for no other reason than because of where they were born.

Rice University professor Scott Solomon[215] is someone looking at the science side of that. The question is, what's going to happen to the first Martian settlers and, more interestingly, their babies?

A Breed Apart

Solomon outlined a number of ways — many of them covered in his Ted Talk[216] — about how humans could change, once we successfully colonize Mars.

1. Humans may develop **denser bones** to overcome the effects of Mars' gravity, which is just a third of Earth's. The reduced force could make bones more brittle, which could lead to complications like fractured pelvises during childbirth.
2. The inhabitants of smaller spaces may become **more near-sighted**, as they no longer need to see as far as they would on Earth. Solomon cites cavefish in deep

trenches that have gone blind with no need for vision, and studies that show children who spend more time indoors are likely to become more near-sighted.

3. Mars inhabitants may develop a **new skin tone** to adjust to the higher levels of radiation. Humans use melanin to fight against ultraviolet rays, while other species use carotenoids. Mars residents may someday have to develop another pigment entirely to fight off radiation.

4. Residents may perhaps learn to **use oxygen more efficiently**. A similar change has been observed on the Tibetan plateau, where oxygen is 40 percent lower than it is at sea level. To adjust, Tibetans have denser beds of capillaries to more efficiently move blood and have the ability to dilate their vessels to get more oxygen to the muscles.

5. One change that could occur relatively fast, is that non-Earth dwelling humans may quickly **lose their immune system**. In a sterile environment with no micro-organisms present, the residents may have no need for a body capable of fighting germs. However, this may not be such a bad thing. Solomon suggests that it could be an opportunity to eradicate diseases, treating the ship flying to Mars as a sort of quarantine zone and ensuring that the new inhabitants can lead healthier lives.

Evolution takes forever, right?

On earth the rate of human evolution is fairly slow, with babies born with 20-120 new genetic mutations. However, throughout history, evolution is faster or slower, depending on how much of an advantage certain mutations afford their owners in making babies. If a mutation gives a person living on Mars a 50-percent survival advantage, it means that those individuals are going to be passing those genes on at a much higher rate than their neighbors otherwise will.

On Mars (due to the increased radiation and generally harsh

environment), Solomon theorizes that the evolution rate will increase dramatically. Instead of waiting thousands of years to witness minuscule changes, Solomon instead believes that humans going to Mars could be on the verge of an evolution-ary roller-coaster. He suggests that some of the changes listed above, could occur within the first few generations, irrevoc-ably splitting the human race.

This roller-coaster, especially if it includes the end of a functional immune system, could result in sex between Martian humans and Earthlings becoming lethal. That could impose an artificial limit on how the two populations will be able to interact and co-mingle. The inability to form families or send offspring back and forth between the two planets, could drive the two groups even further apart.

What about other space settlements?

While Solomon almost exclusively focuses on Martian settlements, many of his postulations would hold for orbital or lunar settlements. Their low gravity, small spaces, high radiation and sterile environments would combine to yield many similar results.

Not This Kind of Doctor

Doctor Who[217] isn't exactly the best example of a space doctor, at least not the kind we are going to need here in the next hundred years. We will need a way to have a doctor with us wherever we go. How can we do that if there are small groups of humans spread across hundreds of stations and millions of miles apart? It isn't exactly feasible to have every 50th person be a doctor. We will examine a couple of solutions to this problem that are rooted in currently available technology.

The Fun Part

June 1, 2063
Swift Crater Settlement, Deimos, 9,300 km from the Martian surface

Greg wasn't thrilled about this. After all, this was his daughter's life. Why should he trust it to a glorified autopilot system? Wasn't there another way?

When Greg was a kid back on Earth, his grandma had a heart attack. Due to complications he wasn't old enough to understand at the time; she had to have surgery with her doctor remotely performing the procedure.

Why couldn't Emma just have that done?

There were doctors on both the Gateway[218] in Lunar orbit, and on the Aldrin Cyclers.[219] Why couldn't they do the surgery remotely?

All of these automated systems were scary. Weren't autonomous cars still having accidents and killing people back on Earth?[220] It wasn't like these machines were perfect. What happens if something goes wrong?

Why did Aquafina not think it was important to have a doctor at the water extraction plant nearby? This was a company settlement after all. Shouldn't the company have brought along a doctor, instead of this jumble of robot limbs? Surely, it couldn't have been cheaper.

Greg had been told about an executive travelling to Mars on one of the cyclers who was evacuated by medical transport back to Earth for his treatment. How could he get that but Emma was stuck with this thing?

Maybe he was worrying too much after all. There wasn't really any other option for her. A medical transport might take six months to get her back to Earth. From what the remote diagnosis had said, her internal bleeding was really bad.

Maybe he should just trust the machine. Maybe he could just rest while she was in there. Maybe, but probably not.

The Real Deal

Greg is going through a problem that we are all going to face, at some point, when we get to space. Where do we get medical care? One solution is to just use the tele-medicine that is starting to become feasible here on Earth.[221] However, that isn't going to work for precise procedures, like surgery. It will probably be fine for routine procedures, but not for the sort of things that require delicate maneuvering, or real-time feedback.[222]

Let us look at what the problem is with using tele-medicine, and possible solutions that we are likely to implement over the next 100 years in space.

Tele-Surgery's Downside

According to medical researchers, in 2018, the ideal amount of latency between doctor and patient during a tele-surgery procedure is 100 milliseconds, and not more than 300 milliseconds.[223] With light traveling 300 km in a millisecond,[224] that means a doctor needs to be within 30,000 km (~19,800 miles) and 90,000 km (~60,000 miles).

What this means for our purposes, is that you have to be really close, at least in space terms. There are currently satellites in geosynchronous orbits that would be in the middle of the safety band.[225] That assumes there is literally no lag on the part of the computer.

Figure 4: A lot of remote surgery now is done with doctors in the same room, rather than separated by thousands of miles

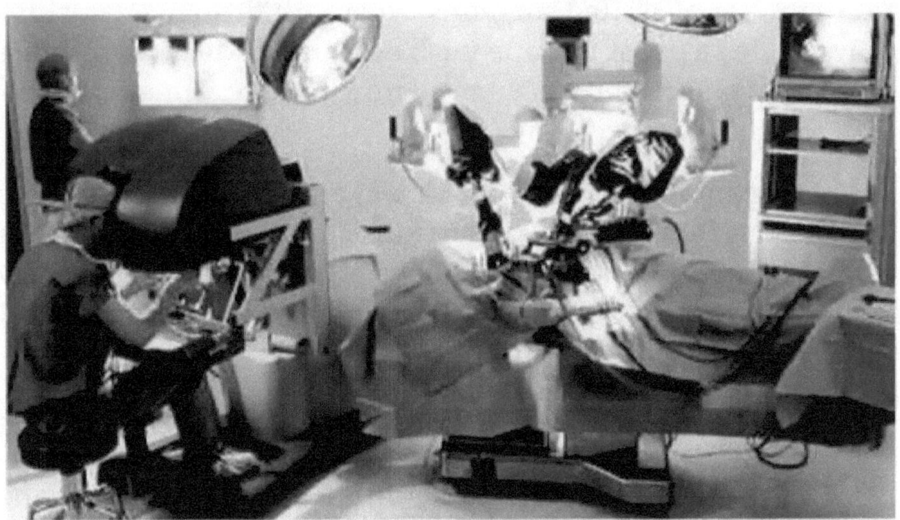

For medical procedures, the distance the signal has to travel is going to be less of a factor, than the ability of the systems on both ends to translate the intentions of a doctor into action.

What Does This Mean?

In practical terms, this means ships traveling between stations, planets, asteroids and just about anything else, will have to carry their own medical support. Stations in close proximity, however, will likely be able to share doctors, as long as there is adequate medical facilities on each for the patient to actually undergo surgery.

The shift towards telesurgery and telemedicine on earth is now allowing relatively scarce doctors to perform risky and delicate procedures for longer, and from greater distances. Without a change in our ability to communicate (e.g. being able to communicate faster than light), we are going to be forced to move beyond current telemedicine ideas.

So, what will we do?

Therefore, if we are limited by the speed of light in how much a surgeon can do remotely, and if we are currently limited in our number of skilled surgeons, what will future generations have to do in order to cope with needing medical care, while they are days or even weeks away from an actual medical facility?

I see three possible options. However, I am sure that there are others (I am open to suggestions/comments). The options are listed below based on their feasibility (based on currently available technology, and our currently accepted medical practices).

Solution 1: Semi-Autonomous Surgeons (Also Known As Robots With Scalpels)

The first, and most likely solution to our space doctor and distance problem, is to take current tele-surgery systems and make them just a bit more autonomous.[226] Think about the difference between your vacuum and my Roomba. Instead of a doctor needing to actually operate the remote surgical system in real time, he can have his robot avatar do pre-programmed procedures based on the patient's needs. This may look like the doctor arranging a sequence of tasks for the robot hands to undertake, with safety parameters that automatically kick in (kind of like how a Roomba will stop rather than run itself down the stairs).

This system requires basic medical facilities on most human outposts/ships, which can receive commands from whatever doctor is needed. It also necessitates at least a semi-skilled human on each ship/station,[227] who can stabilize emergency cases long enough for a signal to get to a doctor and begin treatment. However, with emergency medical technician training being a matter of weeks, compared to at least a decade for a surgeon, that seems like a manageable hurdle.

This system would allow skilled doctors to take care the maximum number of patients across the furthest distance, without sacrificing the 'human in the loop'. Its main downside is that it will take the longest to get a patient treated. The time may be less than the current average ambulance ride. however, it will still take the longest of the three possible future systems.

Solution 2: Chirurgus A Machina (Also Known as a Benevolent Terminator)

Rather than God being in the machine, we could put a doctor in the machine. It may seem a tad far-fetched. However, let me explain.

Surely, We Can Make A Nice Version

First, the human body is just a machine. A complicated one... but a machine, nonetheless. With that being the case, we can remove the human in the loop of the semi-autonomous doctors described above. It would just be a matter of training an algorithm with enough data for it to learn all the ways things can go wrong, and ways to mitigate those problems.[228] This is very similar to how medical students are now taught. There is still a degree of art to medicine, but much of the training involves showing medical students the different ways things go wrong, and the best ways to fix them.

This would be a resource-intensive project up front (making self-driving cars might seem like an easy challenge compared to this), but its main benefit is the ability to 'mass produce' skilled surgeons. The ability to install this Chirurgus (surgeon) system in every ship or station would enable rapid emergency or non-emergency care for all the squishy monkeys living there.

An obvious downside is the lack of a human in the middle, who can reassure us that this robot won't decide to freeze up at exactly the wrong moment. There are also risks to these machines being hacked, or otherwise re-purposed for nefarious aims.

Solution 3: Sleep It Off

The final possibility is that we just freeze patients, until they can reach a doctor. This would be a bit like the stasis[229] pods of just about every science fiction story, but its analogue is currently in use on Earth. We already induce 'therapeutic hypothermia' in patients to prevent brain damage in severe heart attacks.[230] There are currently limitations to how long this can be sustained before it damages the patient, but these hurdles can be overcome.

This system would essentially be a merger of current life support systems (which can keep people alive for as long as 2 years[231]), with a bed where a patient could be chilled from both the inside and outside.
Chilled liquids would run through IVs into the patient,[232] while the bed itself would cool the patient to a 'storage' level.[233] The system would need an integrated ventilator or

other way to deliver oxygen, and ensure that blood continues to circulate.

This whole system would then be shipped to a medical facility, by either the space equivalent of UPS or some sort of specialized medical shipping company, allowing human doctors to revive and repair the patient.

The key benefit of this system is it avoids robots all together (assuming you view that as a benefit), and allows for a centralization of medical providers at a relatively small risk to patients. Its main drawback is that it could take literally years for a patient to get treatment, depending on how long it takes to get back to a doctor.

Which One Is Most Likely?

Because we are silly monkeys at our core, some variant of all three of these will almost certainly see use in space over the next hundred years. We are almost certain to avoid standardizing medical care to one easy-to-use / one-size-fits-all system. The richest among us may opt for specialized, frozen transport to an actual in-person doctor; while automated surgeons provide care to impoverished miners or hospitality workers.

It will depend on our relationship with automation, and the willingness of the people who venture into the stars to recognize the risks of allowing a robot to operate on us isn't greater than the risk of even the best surgeon making a mistake. We fundamentally trust those who look and act more like us, so unless we overcome our instinctive bias against robots, we are more likely to rely on systems that allow for humans in the loop.

Messy Business

Blood is pretty standard down here. It moves from inside to outside, if your skin is opened up. Once outside, it tends to pool on the ground. However, what happens if you bleed in orbit?[234] Let's see what happens, when Zach decides to be clumsy.

The Fun Part

February 23, 2033
En-route to Asteroid 101955 Bennu, approximately 126 million miles from Earth

Get him into the trauma pod, Dan. Stop messing around. He is already missing part of his leg. You aren't going to hurt him more by hurrying this up.

Zach, can you hear me? We are going to get you through this. The trauma pod is going to help us stop the bleeding, and we have enough IV fluids here to stabilize you. However, you are going to need to stay calm.

Zach, look at me. Just stay calm. I know it must really hurt but just look at me.

Dan, why do I still see blood floating around? We have about three minutes before it is pulled into the air system and we are all seriously harmed.

Not that I think you have AIDS, Zach. However, I really would prefer to not be breathing your blood.

Dan, hand me the IV pump, once you get the trauma pod pressurized. We are going to need to get Zach to a bed to strap him down. Then we are going to have to figure out the quickest course back to a hospital.

I'll take care of getting the IV going and getting some sedatives in him, so he doesn't mess with his leg…or what is left of it.

I already let the boss know that something went wrong, while Zach was outside repairing the panel after that impact. However, I didn't have a ton of details then. While you are plotting our new course, can you call home and let them know it looks like something out there explosively decompressed and took off Zach's leg from below the knee?

How this guy got back in, is beyond me. However, we are going to be lucky to stabilize him long enough to get back.

The Real Deal

There are a couple of issues going on here. We will address them all. The first issue is bleeding in micro-gravity. Where does the blood go? Second, how will we deal with trauma in space?

A Vampire's Paradise

For those of you who have never cut yourself bad enough to bleed here on Earth, here is a lesson on what happens:

1. The stuff on the inside can get to the outside. This is mainly the red stuff. However, depending on where you are cut and how deep it is, there can be other stuff too.
2. Your heart pumping (assuming it still is), pushes the blood out of your body.
3. The blood runs from wherever it is pushed, out to wherever the ground is.
4. The end.

In zero- or micro-gravity (this is important because if you are under acceleration, you will have the quasi- gravity that makes wherever the engine is the 'floor'), steps 1-2 are pretty much the same. This is because for no other reason, your body is still doing about the same stuff. Step 3 is where things change, because there is no 'ground'. Instead of your blood running anywhere, it just starts to_pool in place.[235] Before you think that this could be good (the blood forming into a little ball around the wound to help itself clot), remember step 2 is still happening.

The fact that blood is pooling around the wound, just means you can't see the actual wound. The blood is still coming out.

If you have caught on to something I said in the Fun Part, and are thinking that my explanation is missing something then: (1) good attention to detail and (2) shut the hell up. I'll get to that.

Our little lesson on space cuts is up to step three:

1. The stuff on the inside can get to the outside. This is mainly the red stuff. However, it also depends on where you are cut and how deep. It can be other stuff too.
2. Your heart pumping (assuming it still is), pushes the blood out of your body.
3. Your (now on the outside) blood now becomes path dependent. Possible paths are below based on the actual conditions you are in:
    1. Somewhere with no, or poor air circulation
        1. Blood pools in a bubble outside the cut. It just sits there but grows until what was on the inside is now on the outside.
        2. Theoretically your blood should clot – however it will likely be slower than normal because clotting relies at least somewhat on exposure to the air, and if there is no air around the cut (because of aforementioned bubble of blood), then it doesn't clot.
    3. In a vacuum (e.g. space)
        1. The blood exposed to the vacuum begins to boil off, allowing more blood to take its place on the outside. This continues until complete.
    4. Somewhere with robust air circulation

1. Blood is pushed away from your body by airflow. It gets pulled into air vents to be recirculated. This continues until complete.

5. The end

Many people have talked about blood boiling in space, and how much fun that might be. Therefore, let's just skip step 3.2, and instead focus now on 3.3.

Not sure the air filter is enough for this

Over the past two decades of the operation of the International Space Station, we have learned to ensure there is plenty of airflow throughout the station.[236] This is because without it, pockets of exhaled air just stay in front of your face. This can suffocate you, if you don't move around on your own.[237] Ensuring that there is plenty of airflow may end up being a bit of a problem in a trauma situation. This is because it may be enough air moving to make the blood move as well. And the blood, like any good liquid, is just going to follow the current, right into the re-circulation system.

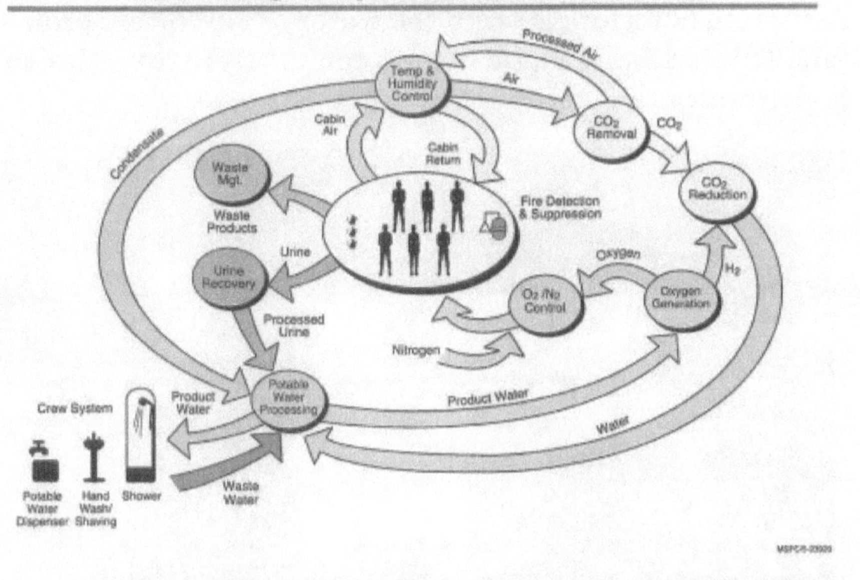

ISS Life Support System[238]

Not sure if you caught it, but I didn't say air re-circulation system. The air system is connected to the water system. Therefore, your little blood droplets are moving along with the rest of the air into the re-circulation system (as depicted in the totally readable and not at all complicated picture above). What happens next?

Much of the liquid in the air is pulled out and sent for filtration to be used as potable water.[239] The air itself is recycled, filtered and then sent back into the station.[240]

Because you were clumsy enough to cut yourself and happened to be on a station that had good enough airflow to prevent you from suffocating, your blood gets to become both an airborne and waterborne hazard to everyone else on board! Yay! The gift that keeps on giving!

We can probably assume you don't have any communicable diseases (I am polite like that), so then what is the risk to everyone else?

Assuming not a lot gets into the water or air you are probably safe. This is despite blood being (technically) a toxin due to its high iron content.[241]

Don't want to drink/breathe your buddy's blood?

A solution to this problem, is to build filtration systems that are able to:

1. Remove toxins from blood (ranging from iron to germs)
2. Ensure that all filtration systems (water and air) have ultraviolet sterilization systems built in to kill those germs that get through the filters.
3. Remember that you are in space and already drink-

ing everyone else's urine, so maybe you should be less squeamish.

Another solution is to just not be clumsy and get hurt. I call that plan A. However, I can't speak for you.

WILL IT BE UTOPIA?

We don't usually think about the Coast Guard whenever we take a cruise or get on a boat, but their infrastructure ensures that no matter where in the U.S. we get on an ocean-going vessel, we are safe. We are at least safer than we would be. Maybe it is time to develop something like that in space. It certainly seems like more of a near-term need than a *Starship Trooperesque* Space Force.[242] Ariana is about to discover what a key task of this Space Guard would have to be. Let's check in with her.

The Fun Part

She was amazing, Ariana thought. How had less than 100 people built something like this in less time than it took her to finish her graduate degree in Orbital Sociology? Ariana was looking at a replica of her new ship, the *USS Endeavor*, the first of a new class of space cutters, while marveling at how real the

$1/3^{rd}$ earth gravity felt in the spinning central hub of the ship.

Ariana assumed command of the ship after her time at the Space Service Command and Staff College. The Endeavor, at nearly 175 meters long, was nearly three times the size of her space shuttle namesake. Endeavor's central shaft hosted her storage, fuel, and shuttles, while the crew lived and worked almost exclusively in the nearly 100-meter rotating central hub. The fold-able solar arrays extended another several hundred meters in each direction, giving the ship, which had a nuclear reactor, the possibility of operating on solar power, even when the ship was within the orbit of Pluto.

Ariana was snapped out of her reverie by a chime on her desk computer. Answering it, she saw the face of her ruggedly handsome executive officer, "What's up Tim?" she said, banishing any awkward thoughts. "Ma'am, we just received orders from HQ, we are to respond to a distress call from ERO 9". Ariana quickly thought back to her briefing on the elevator ride skyward, ERO, or Easily Relocatable Objects were asteroids in heliocentric orbit, which were fairly easy to move into lunar or earth orbits. ERO 9 was a Canadian claimed rock nearly 1 kilometer in diameter, and its rare minerals were worth between 1 and 3 trillion dollars.

"What could those Canadians have gotten into?" She responded. Tim smiled ruefully, "Looks like their water recycler has been having problems for a few weeks, and they 'forgot' to order spare parts." "Looks like a repeat of ERO 2 when the Chinese administrator was scared to report the problem, and the whole team died of dehydration."

"Damn it Tim! That is the last thing that we need now. Is the Endeavor even ready to fly?" "Yes Ma'am, although best case we are six days out." "Exactly Tim, now it is our fault that we can't save the stupid idiots who couldn't be bothered to keep up with their maintenance" "Looks like it is Ma'am…but isn't that why they pay us the big bucks?

"Tim, remind me again why I joined the Coast Guard equivalent? Don't answer that you snarky jerk. – I know it is better than flying a desk back at the Pentagon…let's just get this done."

The Real Deal
  The U.S. Space Guard

As early as 2000 researchers were conceptualizing a consolidation of all commercial launch-to-orbit-to-landing oversight and regulatory functions within the "prevention" arm of a new "Space Guard."[243] This narrative largely went unheeded,[244] after significant resistance from the Defense Department and NASA, both of whom saw the new entity as a threat to their roles in space.

After nearly 20 years, a new space service has begun to 'take off' (haha, get it?). However, much of the commentary on the creation of a Space Force presumed that it would be an Air Force version of the Marine Corps.

Maybe, but is that really the best idea? Does a space "Coast Guard" make more sense? Depending on the core functions it performs, maybe Space Force should be a civilian agency.

Looking into the distant future, establishing the Space Force as a civilian agency would also future-proof the organization against legal prohibitions on the establishment of military installations on celestial bodies.[245] We may not need a space navy or starship troopers in any of our lifetimes. However, we almost certainly will have permanent commercial facilities on the moon or various asteroids. This creates the need for a commensurate civil governmental presence.

How Would the Space Guard Work?[246]

Nearly every authority that the United States needs for effective maritime governance is vested within the Coast Guard,[247] and the service's 11 statutory missions[248] cover the full panoply of possible action within the entire maritime domain. This means that the Coast Guard is the lead U.S. federal agency for nearly every matter that takes place within the navigable waters of the United States or that pertains to U.S. vessels or vessels in which the United States may exert jurisdiction.

The agency also has wide ranging authority to coordinate and cooperate with other Federal and state agencies and foreign governments, conducting maritime mobility operations, like ice-breaking, and aid-to-navigation placement and maintenance. Its most well-known mandate though is to "perform any and all acts necessary to rescue and aid persons and protect and save property on and under the high seas and on and under water over with the United States has jurisdiction".

Applying this construct to a similarly organized and empowered Space Guard would concentrate space expertise within one agency, lowering regulatory cost burdens on the U.S. private space industry. It would also lower the risk for companies and countries, as they push further from near Earth orbit. Instead of having to assume all the security and safety risk, some of that can be outsourced to this new government entity.

The first problem that this new Space Guard could work on as a 'quick win,' would be space debris. There are currently tens of thousands of pieces of space debris, and no entity with the authority or capability to clean it. There are hundreds, if not thousands, of satellites beyond their expected life, just waiting to be de-orbited.[249] The organization to manage the 'cleaning' of orbit would do more than just free up space for new satellites. It would create the framework for international cooperation in space.

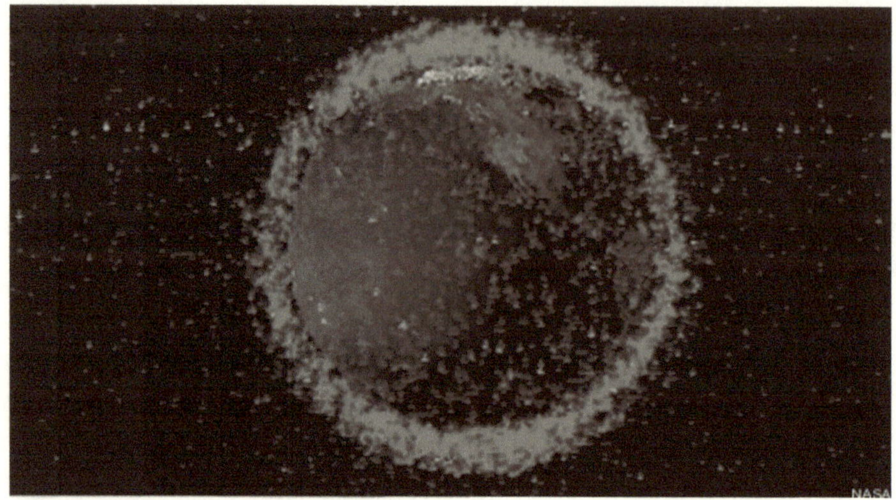

A 3D representation of the location of all known space debris.

Water Recycling

I know it sounds weird that a water recycling system failing would constitute an emergency. It is not so much us needing water that is weird, but more the thought that there wouldn't be extra water handy. However, as we get more and more comfortable in space, our 'acceptable' minimum amount of emergency water will likely decrease. This combined with the fact that water is heavy, and takes up a lot of space, means bringing it up to orbital installations isn't ideal.

Recycling water onboard stations is already done at the International Space Station.[250] It will likely be a fixture in future space stations. They will likely rely on semi-regular deliveries of water (maybe as little as once or twice a year) to replenish their systems. However, they will otherwise be self-sufficient.

There is an entire closed-loop system onboard the International Space Station that is dedicated to water. First, astronaut wastewater (such as urine, sweat, or even the moisture from their breath) is captured. The impurities and contaminants are then filtered out of the water. The final product is potable water that can be used to re-hydrate food, bathe, or drink.

Repeat. The system sounds disgusting, but recycled water on the ISS is cleaner than what most of us Earthlings drink.

How this works is a little more complicated. Even the European Space Agency's public affairs team[251] needed two infographics to portray it all. You are welcome to try and trace all of the lines in the graphics below, but I didn't have any luck. All I learned was there was about a score of places where a lack of maintenance could cause the whole system to break down.

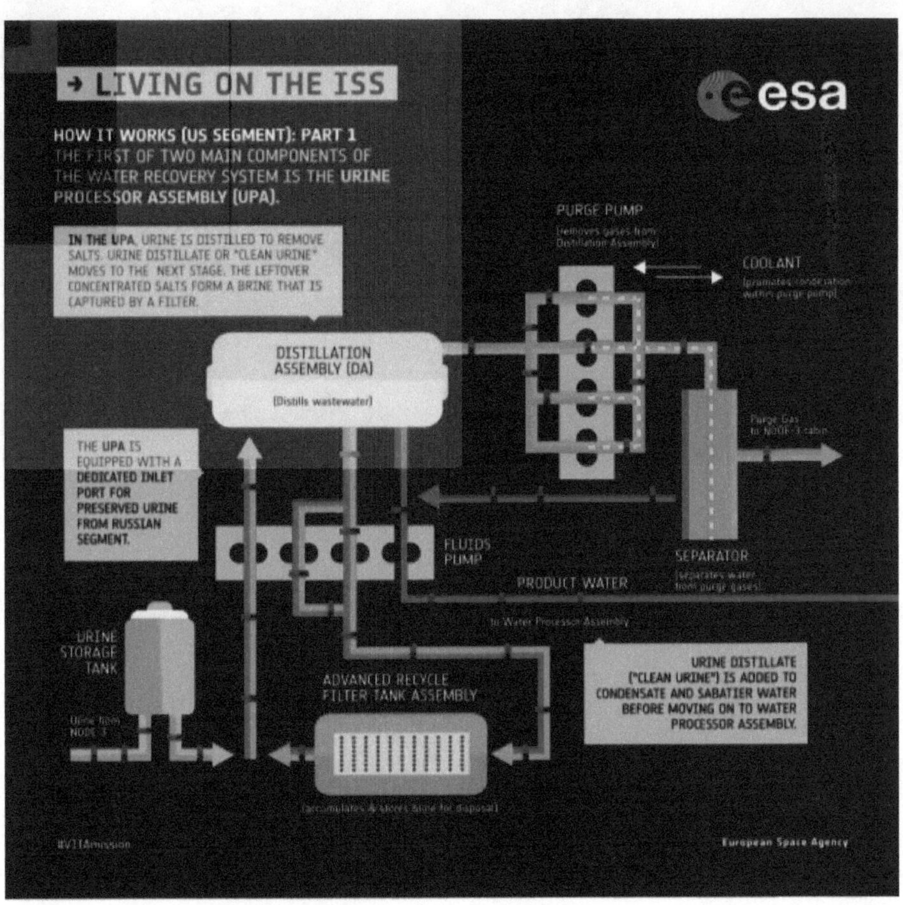

The International Space Station's Water Recycling System Part 1[252]

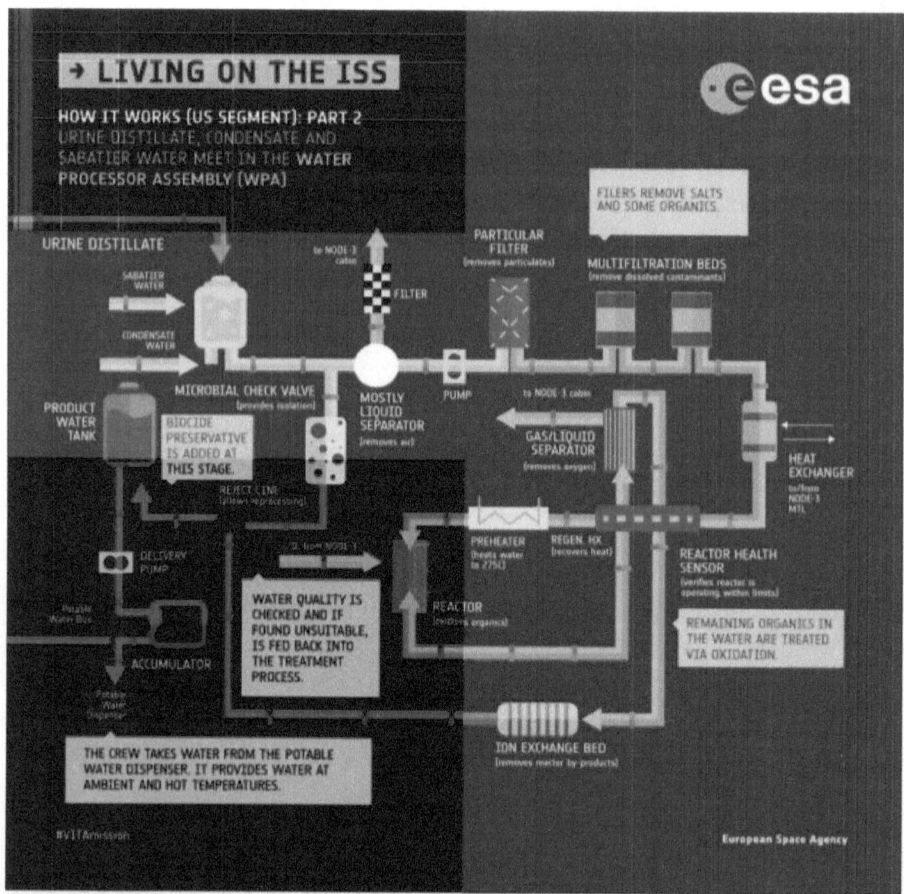

The International Space Station's Water Recycling System Part 2[253]

Dissident Outlet Or Brain Drain?

What happens when more people have the opportunity to immigrate to space? What does it look like, when oppressed communities once again have a new world to escape to? Will this herald a new era of rigidly homogeneous cultures? Or will earthly countries just hemorrhage their best and brightest to these new frontiers? Let's see what it looks like for Kim in one of these futures.

The Fun Part

March 19, 2073
Lunar Landing Zone 4; 95th Enclave

Today is a pretty good day, Kim decided. After all, it was her birthday. Well, not like the day she was evicted from her mother's womb. It was the day that she had first stepped off the lunar cycler shuttle onto the 95th Enclave. The 95th was a Lutheran colony that was only two years old when Kim first arrived. However, it was one of the dozens of niche colonies springing up all over the moon.

Today, Kim and the 32 other Lutherans who had arrived with her, would have their true birthday party. It was one of the twice-monthly ceremonies held for whomever had arrived on one of the two lunar cyclers[254] rotating between Earth and the moon.

It was the only two days a month when everyone got actual fresh food for the whole meal, instead of that vat grown meat and nutrient paste. It really was a big deal Kim thought. When she was growing up, her parents treated her day of baptism as just as important as her 'real' birthday. Therefore, doing something similar here didn't seem that odd.

What was a little odd, was how quickly the Lutheran church back on earth seemed to be dying. When Kim left just two years ago, there were almost a dozen Lutheran churches in her hometown of Minneapolis. However, now there were only three. That seemed to make the sense of urgency to get the 95th ready to support other faithful that much more potent.

After all, if the church was going to be persecuted into oblivion back on Earth, then it was up to the faithful to prepare a safe place for those chosen of God. One day, God willing, the church could return to its birthplace on Earth, and spread the gospel to the lost and dying.

The Real Deal

OK, a little strange right? Why is there a Lutheran colony on the moon? Why am I suggesting that the Lutherans are going to be persecuted in America sometime in the next 60 years? It has nothing to do with the actual Lutheran denomination of Christianity. They are just an easy thing to write about, because there is a Lutheran church across the street from where my kids live.

Second, the point is that there is always going to be a group, whether religious, ethnic, ideological or something else, that is outside of the mainstream. Regardless of how accommodating the mainstream is, the outsiders will feel they are in danger of disappearing. This becomes especially true, if the outsiders are being repressed.

For one of the first times in history, there isn't anywhere on Earth for these groups to resettle, if their home doesn't want them. There is no frontier, colony or 'extra' space for these people to be put, to get them out of sight and mind. As the cost to get to orbit decreases, this dynamic will shift. All of a sudden, the most effective punishment in history (banishment)[255] becomes a viable option for governments, organizations and groups.

These groups will also have somewhere to go on their own, if they want to maintain their culture, customs, beliefs and/or practices. This has the potential to create a future where conflict between disparate groups actually decreases, as the most hardline believers move to homogeneous off-world enclaves.

Historical precedent?

One historical precedent for this possible future is the Puritan emigration from England to New England in the 1600s.[256] This 'great migration' provided both the foundation of New England's culture, and eliminated a large, if not the largest opposition group to mainstream Catholicism/Protestantism in England.

Most of the Puritans came from the eastern counties of England. They tended to be tradesmen or skilled craftsmen rather than farmers, since tradesmen and craftsmen tended to be more highly educated than was usual for the time. They were also affluent enough to be able to afford to pay for their own passage, and migrated in small, nuclear families, or as single men.

Once in the New World, the Puritans ensured complete adherence to their ideology through an aggressive enforcement system, which saw offenders banished. In many cases, the system the Puritans set up in their new colonies was more oppressive than the ones from which they fled.

Nevertheless, there have been some who credit the migration of groups, like the Puritans, with stabilizing England and the European continent, after generations of religiously motivated wars. No longer were rival religious groups competing for land and resources, allowing governments to consolidate their power, permitting the ideas of the Enlightenment to spread, and ultimately setting the conditions for the first Industrial Reformation, transforming the world.

Yes, but what about space?

For at least the next 50 years, the people who go to space will have to be skilled enough to improvise, adapt and innovate just to survive. This new frontier will be less Survivor,[257] and more Naked and Afraid.[258] As a result, we are likely to see emigration from Earth, which looks similar to what happened with the Puritans.

There will no doubt be a swift movement of highly educated, intelligent and skilled people to new colonies. However, it is unlikely that all the smarties are headed away from earth. Indeed, it is more likely that oppressed communities, and the governments doing the oppressing will view space as a tantalizing opportunity. No longer would China have to expend the resources to monitor and reform millions of dissidents,[259] they could just be exiled to an easily manageable asteroid, or other such outpost.

Similarly, white nationalists, Antifa, or even people who just want Game of Thrones to have a different ending, will be free to create their version of paradise on these new worlds. The smartest, most affluent and skilled of these groups are likely to pioneer new, utopian colonies, simultaneously depriving earthbound dissidents of leadership, and opening new opportunities for these groups to flourish.

Adam's Bad Day

April 3, 2076
Geosynchronous Earth Orbit, somewhere over California

Adam's life was anything but exciting. Having dreamed of joining the Space Force (now the newly independent Space Corps), he thought he would be living the high life by now. Literally. He had gone to the right schools, earned the right degree (orbital topography) and been at the top of his Space ROTC class. However, he was somehow assigned to lead a repair team. A repair team! Sigh.

Well, he thought, it was a little better than that – he was technically the flight commander for sixteen repair teams responsible for the maintenance of the aging Astro-Particle Beam System...or APBS for short. It wasn't so much that he hated the work. It was interesting enough. The APBS were mainly antiques after all, the first one launched in 2022, and the newest was about 25 years old. Adam also had a soft spot in his heart for the classics. It was why his troopers always teased him about his Justin Timberlake playlist.

No, it wasn't the work. It was the tedium and the pointlessness of it all. These systems had been launched at the height of tensions during the Sino-American conflict, and had been instrumental in removing nuclear weapons from the table during the Chinese invasion of Taiwan, the reunification of North and South Korea and the establishment of the new Japanese empire in the Pacific. They were useful then, but it had been decades since anyone even thought about using nuclear weapons, especially with all of the defense money going towards competition between the great powers on Luna and Mars. He could have been out there – his ex from SROTC was on Luna managing one of the defense stations there. Now, that would have been interesting.

Just as he turned his attention back to checking the work his epsilon team had done on APBS-6, he noticed a meteor storm on the horizon. "Was that over Europe?" he thought absently. As he swiveled back to his work, his mind nagged at him to look back. Something wasn't right about the meteors he just saw.

Before he could turn, he heard "ALL HANDS BRACE, BRACE, BRACE....THIS IS NOT A DRILL" over his headset.

Grabbing for the nearest pylon, Adam began cycling through his radio channels trying to raise his commander. As he did, APBS-6 began to pivot, until Adam could see the meteor storm he noticed earlier. However, the meteors weren't plunging towards earth, they were moving up.

Cold horror crept into his bones, as Adam started to think that maybe he shouldn't have wished for his life to get more interesting. As the system vibrated beneath his boots, Adam laughed uncontrollably. He had been one of the people who had sworn blindly that the system was not only not needed, but also that the physics of the system meant they would be silent and motionless. Now, as the system fired for the first time, Adam couldn't think of anything except how wrong he had been. Well, that and how he hoped everything he thought he knew about nuclear explosions in space was accurate. Otherwise, he was about to be adrift in his unshielded repair tug.

The Real Deal

"Peaceful use of space".

That has always been a pipe dream, talked about by politicians and pined for by academics. The first use of space has always been to achieve military dominance, or at least parity with one's enemies. Without a significant shift in geopolitics between now and the end of the 21st century, the story described above is extremely likely.

The U.S., China and Russia are already exploring launching space-based defensive systems to be able to thwart their opponent's new (or just about to emerge from development) hyper-sonic missiles.[260] This may be good in the short run, since these systems prevent any single power from gaining too much power over another. However, if history is a guide, they also mean it is only a matter of time before their limits are tested by a country desperate enough to risk everything for what seems like their survival.

One day we will realize we are all stuck on this planet together, and there isn't another one viable right now. Therefore, we should probably avoid cracking the mantle open with projectiles in the hope of getting ahead.

A Shield for Whom?

Zehra joined the Shield of Iron private security firm in early 2055, leveraging her time in the Turkish military to land a coveted position in Shield's orbital security services (SOSS). They were charged with protecting a joint French and German space-based solar power plant.

The work wasn't exactly glamorous, but with it paying three times what she would make groundside, there wasn't much thought of returning to Earth, other than for her paid leave. The pay, and the family care program SOSS offered, in hindsight, were probably why Zehra never raised an objection when she received that communication early one Thursday.

That Thursday would have been memorable anyway, because it was Zehra's sister's 18th birthday. The call from the SOSS HQ in Istanbul came about 10 minutes after Zehra finished wishing her sister well. Yusuf, the chief of operations at SOSS, told her to seize the VIP transport shuttle, which was passing her orbital facility in 30 minutes. While Yusuf made an effort to explain that the transport was carrying a wanted fugitive and Zehra and her crew would receive a substantial bounty, Zehra never even thought to question the order. Within 10 minutes, three men, two women and Zehra were on board their fast patrol cutter and burning towards the shuttle.

Within an hour, and after two of her team were injured in the boarding, Zehra reported to Yusuf that the shuttle was secure, but everyone on the shuttle had been killed. Zehra started to ask Yusef if that would be a problem, when the shuttle's sensors detected the first of what would be a dozen hyper-velocity missiles launched from a Japanese ground installation targeting the ship that had raided the

Japanese Deputy Prime Minister.

The Real Deal
Space-Based Solar Power (SBSP)

The idea of capturing solar power in space for use as energy on Earth, has been around since the beginning of the space age. In the last few years, however, scientists around the globe — and several researchers at the US Energy Department's Lawrence Livermore National Laboratory (LLNL)[261] — have shown how technological developments could make this concept a reality.[262] In early 2018, scientists from the California Institute of Technology announced[263] that they had succeeded in creating a prototype capable of harnessing and transmitting solar energy from space.

Their prototype is a lightweight tile that consists of three main components. Optical reflectors concentrate the sunlight, photovoltaic cells convert the sunlight to electricity and an integrated circuit converts the electricity to radiofrequency energy that is transmitted through an attached antenna. Many individual tiles could be strung together to form large solar arrays in space. A ground-based microwave receiver on Earth would be used to intercept the incoming radiofrequency energy and convert it back into usable electricity.

The scientists demonstrated that their prototype works, by subjecting it to space-like conditions in the laboratory and using it to power a light-emitting diode (LED) located about 20 inches (50 centimeters) away from the tile. You can view the prototype in action at the 2:26 mark in the video[264] at the link here.[265]

Advantages of SBSP?[266]

Solar arrays, no matter how efficient, only generate power

when the sun is illuminating them. Therefore, if solar power is going to replace today's fossil-fuel-powered electric-generating stations, it must be generating power all the time. This is not possible for ground-based solar energy without some form of electrical energy storage, which adds significantly to the cost of solar electricity generated on Earth.

The great thing about space, is that orbits exist where there's no nighttime. We, therefore, have the prospect of making power that flows continuously and that can be instantly sent to where it is needed. The potential benefits are enormous. About a quarter of humanity has no electric power whatsoever. This is an enabling technology that could leapfrog the electric-power transmission grid on Earth, and have the same effect that the cellular phone system had on communications.

Types of SBSP[267]

The two most commonly discussed designs for SBSP are a large, deeper space microwave transmitting satellite and a smaller, nearer laser transmitting satellite.

Microwave Transmitting

Microwave transmitting satellites orbit Earth in geostationary orbit (GEO), about 35,000 km above Earth's surface. Designs for microwave transmitting satellites are massive, with solar reflectors spanning up to 3 km and weighing over 80,000 metric tons. They would be capable of generating multiple gigawatts of power, enough to power a major U.S. city.

The long wavelength of the microwave requires a long antenna and allows power to be beamed through the Earth's atmosphere, rain or shine, at safe, low intensity levels hardly stronger than the midday sun. Birds and planes wouldn't notice much of anything flying across their paths.

The estimated cost of launching, assembling and operating a microwave-equipped GEO satellite is in the tens of billions of dollars, based on current launch costs of ~ $1,000 USD per KG. It would likely require as many as 40 launches for all necessary materials to reach space. On Earth, the rectenna used for collecting the microwave beam would be anywhere between 3 and 10 km in diameter, a huge area of land and a challenge to purchase and develop.

Laser Transmitting

Laser transmitting satellites, as described by LLNL,[268] orbit in low Earth orbit (LEO) at about 400 km above the Earth's surface. Weighing in in at less than 10 metric tons, this satellite is a fraction of the weight of its microwave counterpart. This design is also cheaper. Some predict that a laser-equipped SBSP satellite would cost nearly $500 million to launch and operate. It would be possible to launch the entire self- assembling satellite in a single rocket, drastically reducing the cost and time for production. In addition, by using a laser transmitter, the beam will only be about two meters in diameter, instead of several km, a drastic and important reduction.

While this satellite is far lighter, cheaper and easier to deploy than its microwave counterpart, serious challenges remain. The idea of high-powered lasers in space could draw on fears of the militarization of space. This challenge could be remedied, by limiting the direction that the laser system could transmit its power.

At its smaller size, there is a correspondingly lower capacity of about 1 to 10 megawatts per satellite. Therefore, this satellite would be best as part of a fleet of similar satellites, used together.

There is also the challenge of these satellites being in LEO, so they would have to constantly be shifting which ground station they delivered power to, as they orbited the entire globe about six-eight times a day.

A New Interstellar Order?
A key consequence of SBSP is the democratization of space. A durable source of limitless power will allow the creation of orbital scale elevators.[269] This will reduce the cost of getting to space from about $10/kg in 2050, to somewhere between one to five cents a kg, once the elevators are complete. For at least the first 20+ years, it is almost certain that the cost of using an orbital elevator will be limited only by capacity, not the cost to operate.

As space becomes increasingly democratized, the stability of our current rule-based international system becomes increasingly perilous.[270] By 2040, a trip to space will cost little more than a trans-Atlantic flight now. Therefore (even without an orbital elevator), any two-bit terrorist group, or just some rich, mischievous kid, will be able to wreak havoc on multi-billion-dollar orbital installations.[271] At some point, countries (and almost certainly companies) will find it cheaper to pay private security companies to secure their orbital installations, rather than field their own space 'forces'.[272]

Who do these private security companies report to? Who governs their actions? Who enforces international rules?[273] There are no boundaries in space, and the thought of drawing lines in space to delineate 'spheres of influence' seems a little silly, when you realize how big space is.

Will force be the only thing that keeps everyone in line in the future? Humanity's past suggests that is the case. However, maybe there is a chance to change that. It could be that realizing you can't see any country's borders in space, will remind future orbital citizens that we really all in this together.

It is a nice thought — and one I hope to be able to help make happen.

CONCLUSION

I hope you enjoyed reading the first pre-history of humanity's second century in space. If you are walking away with the impression that humans are just fancy monkeys who are going to act the same in a jungle as they do in orbit then I did my work correctly.

If you still think that humans are going to be sweet cyborgs plying the interstellar tradeways[274] in peaceful pursuit of knowledge and happiness…well…there is no helping you.

So, go forth, with either your new insight into the future, or the knowledge that literally anyone can write a book, and do humanity proud.

See you out there in the deep black of space, and if you want to learn more about the second century of humanity in space feel free to follow me at www.future.humanityinspace.com or @humanityinspace on Twitter.

-Tim

[1] Wulder, Michael,Thomas Hilker, Joanne White, Nicholas Coops, Jeffrey Masek, Dirk Pflugmacher, Yves Crevier. (2015). Virtual constellations for global terrestrial monitoring. *Remote Sensing of Environment. 70*, 62-76. https://www.sciencedirect.com/science/article/pii/S0034425715301243

[2] Edwards, Bradley C. (March 1, 2003). *The space elevator: NIAC phase II final report*. NASA Institute for Advanced Concepts. http://images.spaceref.com/docs/spaceelevator/521Edwards.pdf;
Raitt, David. (Decemember 2015). A historical look at the concept of space elevators. *New Space, 3*(4), 231–238. https://doi.org/10.1089/space.2015.0024

[3] Pearson, Jerome. (1975). The orbital tower: A spacecraft launcher using the Earth's rotational energy. *Acta Astronautica. 2*, 785-799. http://www.star-tech-inc.com/papers/tower/tower.pdf

[4] ESOA (EMEA Satellite Operators Association). (No date given). *Satellite orbits*. https://www.esoa.net/technology/satellite-orbits.asp

[5] Howell, Elizabeth. (April 24, 2015). *What is a geosynchronous orbit?* Space. https://www.space.com/29222-geosynchronous-orbit.html

[6] Elevator World. (No date given). *The maximum speed of elevators*. https://www.elevatorworld.com/the-maximum-speed-of-elevators/

[7] Phoenix Modular Elevator. (No date given). *Speed does not mean fast*. https://www.phoenixmodularelevator.com/elevator-speed/

[8] Clark, Stephen. (April 4, 2019). *Israel's Beresheet lander brakes into lunar orbit*. Spaceflight Now. https://spaceflightnow.com/2019/04/04/israels-beresheet-lander-brakes-into-lunar-orbit/

[9] Encyclopaedia Britannica. (No date given). *Friction*. https://www.britannica.com/science/friction

[10] Nardi, Tom. (March 18, 2019). *Hitchhiking to the Moon for fun and profit*. Hackaday. https://hackaday.com/2019/03/18/hitchhiking-to-the-moon-for-fun-and-profit/

[11] Debre, Isabel. (April 4, 2019). Israeli spacecraft enters lunar orbit ahead of moon landing. *U.S. News & World Report*. https://www.usnews.com/news/business/articles/2019-04-04/israeli-spacecraft-enters-lunar-orbit-ahead-

of-moon-landing

[12] Smithsonian National Air and Space Museum. (No date given). *Moving in space*. How Things Fly. https://howthingsfly.si.edu/flight-dynamics/moving-space

[13] Scuka, Daniel. (October 17, 2016). *Burn baby, burn! The technology of the Mars orbit insertion burn*. European Space Agency (ESA). https://blogs.esa.int/rocketscience/2016/10/17/burn-baby-burn-the-technology-of-the-mars-orbit-insertion-burn/

[14] Hadhazy, Adam. (December 22, 2014). A new way to reach Mars safely, anytime and on the cheap. *Scientific American*. https://www.scientificamerican.com/article/a-new-way-to-reach-mars-safely-anytime-and-on-the-cheap/

[15] Smithsonian National Air and Space Museum. (No date given). *Moving in space*. How Things Fly. https://howthingsfly.si.edu/flight-dynamics/moving-space

[16] Smithsonian National Air and Space Museum. (No date given). *Gravity assist or "planet hopping"*. How Things Fly. https://howthingsfly.si.edu/media/gravity-assist-or-planet-hopping

[17] Oberg, James and Buzz Aldrin. (March 2000). A bus between the planets. *Scientific American, 282*(3), 58-60. https://www.scientificamerican.com/article/a-bus-between-the-planets/

[18] Genova, Anthony L. and Buzz Aldrin. (2015). *A free-return Earth-Moon cycler orbit for an interplanetary cruise ship*. Presented at the AAS/AIAA Astrodynamics Specialist Conference. https://ntrs.nasa.gov/archive/nasa/casi.ntrs.nasa.gov/20150018049.pdf

[19] Aldrin, Buzz. (No date given). *Aldrin Mars cycler*. https://buzzaldrin.com/space-vision/rocket_science/aldrin-mars-cycler/

[20] Dorling Kindersley, Limited. (2016). Living on Mars. In *How Super Cool Stuff Works* (pp.142-143) Dorling Kindersley, Limited.;
Mars One. (No date given). *How long does it take to travel to Mars?* https://www.mars-one.com/faq/mission-to-mars/how-long-does-it-take-to-travel-to-mars

[21] Aldrin, Buzz. (Oct 28, 1985). *Cyclic trajectory concepts.* Science Applications International Corporation (SAIC) presentation to the Interplanetary Rapid Transit Study Meeting, Jet Propulsion Laboratory. https://web.archive.org/web/20180731213226/http://buzzaldrin.com/files/

pdf/1985.10.28.ALDRIN_SAIC_PAPER.Cyclic_Trajectory_Concepts_0.pdf
[22]

Aldrin, Buzz, Dennis Byrnes, Ron Jones, and Hubert Davis. (2001). *Evolutionary space transportation plan for Mars cycling concepts*, Paper AIAA 2001-4677 presented at the American Institute of Aeronautics & Astronautics Conference and Exposition. https://arc.aiaa.org/doi/10.2514/6.2001-4677

[23] Aldrin, Buzz and Leonard David. (2015). *Mission to Mars: My vision for space exploration*. National Geographic Books.

[24] Nock, Kerry T. (2001). *Frequent, fast trips to and from Mars via astrotels.* Global Aerospace Corporation. http://www.gaerospace.com/projects/AstroTels/pdfs_docs/SSI_HF_Astrotel_Nock.pdf;
Radford, Tim. (February 7, 2002). Space hotels could run ferry service to Mars. *The Guardian.* https://www.theguardian.com/world/2002/feb/07/physicalsciences.spaceexploration

[25] Nock, Kerry T. (June 7, 2000). *Cyclical visits to Mars via astronaut hotels.* Presentation to the NASA Institute for Advanced Concepts (NIAC) 2000 Annual Meeting. [PowerPoint slides]. SlideShare. https://www.slideshare.net/cliffordstone/visits-tomars

[26] Jarvis, Michaela. (May 4, 2020). *Is bacteria the key to growing food in space? Eagles see potential.* Embry-Riddle Aeronautical University News. https://news.erau.edu/headlines/is-bacteria-the-key-to-growing-food-in-space

[27] American Chemical Society [press release]. (April 2, 2019). *Bacterial factories could manufacture high-performance proteins for space missions.* https://www.acs.org/content/acs/en/pressroom/newsreleases/2019/april/bacterial-factories-could-manufacture-high-performance-proteins-for-space-missions.html

[28] Mathewson, Samantha. (September 25, 2019). *The Gut in Space: How Bacteria Change in Astronauts' Digestive Systems.* Space. https://www.space.com/astronaut-gut-bacteria-changes-in-space.html

[29] The Guardian. (No date given). *Heston Blumenthal.* https://www.theguardian.com/food/heston-blumenthal

[30] Best Restaurants of Australia. (No date given). *Heston Blumenthal's profile.* https://www.bestrestaurants.com.au/chef/heston-blumenthal/profile

[31] Lamont, Tom. (March 5, 2016). Heston, we have a problem... the top chef cooks for Tim Peake. *The Guardian.* https://www.theguardian.com/lifeandstyle/2016/mar/05/heston-blumenthal-chef-cooks-astronaut-tim-

peake

[32] NASA. (October 23, 2010). *Flight engineer Sandra Magnus' Space Station journals.* https://www.nasa.gov/mission_pages/station/expeditions/expedition18/journals_sandra_magnus.html

[33] Magnus, Sandra. (October 23, 2010). *Sandra Magnus' journal: Food and cooking in space.* NASA. https://www.nasa.gov/mission_pages/station/expeditions/expedition18/journal_sandra_magnus_6.html; Magnus, Sandra. (October 23, 2010). *Food and cooking in space, part 2.* NASA. https://www.nasa.gov/mission_pages/station/expeditions/expedition18/journal_sandra_magnus_7.html

[34] Misra, Ria. (April 24, 2014). *A complete guide to cooking in space.* Gizmodo. https://io9.gizmodo.com/what-happens-when-you-cook-french-fries-in-space-1566973977

[35] USDA Food Safety and Inspection Service. (September 2011). *Food safety information: High altitude cooking and food safety.* https://www.fsis.usda.gov/shared/PDF/High_Altitude_Cooking_and_Food_Safety.pdf

[36] NASA. (No date given). *Closing the loop: Recycling water and air in space.* https://www.nasa.gov/pdf/146558main_RecyclingEDA(final)%204_10_06.pdf

[37] National Air and Space Museum. (No date given). *Food In space.* https://airandspace.si.edu/exhibitions/apollo-to-the-moon/online/astronaut-life/food-in-space.cfm

[38] Pappalardo, Bruno. (February 18, 2019). *What did sailors in the Georgian Royal Navy eat?* HistoryHit. https://www.historyhit.com/what-did-sailors-in-the-georgian-royal-navy-eat/

[39] British Food in America. (Winter/Spring 2020). *Food at sea in the age of fighting sail, 63.* https://www.britishfoodinamerica.com/Our-First-Nautical-Number/the-lyrical/Food-at-sea-in-the-age-of-fighting-sail/#.XuGaoEVKjIU

[40] Kiang, Charlotte. (January 16, 2020). *A freshly cooked meal in space? It could happen sooner than you think.* Forbes. https://www.forbes.com/sites/charlottekiang/2020/01/16/a-freshly-cooked-meal-in-space-it-could-happen-sooner-than-you-think/

[41] Foust, Jeff. (March 21, 2019). *Blue Origin studying repurposing of New Glenn upper stages.* SpaceNews. https://spacenews.com/blue-origin-

studying-repurposing-of-new-glenn-upper-stages/

[42] O'Callaghan, Jonathan. (December 11, 2019). Blue Origin launches its first space tourism rocket in seven months - and hopes to take humans to space in 2020. *Forbes.* https://www.forbes.com/sites/jonathanocallaghan/2019/12/11/blue-origin-launches-first-space-tourism-rocket-in-seven-months-ahead-of-planned-human-flights-in-2020/#679c3a5afe6e

[43] https://www.hgtv.com/

[44] Buckner, Brett. (February 7, 2009). Origin of debate. *Anniston Star.* https://www.aarweb.org/common/Uploaded%20files/Awards/2010%20Buckner.pdf

[45] Suddath, Claire. (October 29, 2009). How do countries determine their time zones? *Time.* http://content.time.com/time/world/article/0,8599,1933079,00.html

[46] Wonderopolis. (No date given). *Why do we have different time zones?* https://www.wonderopolis.org/wonder/why-do-we-have-different-time-zones

[47] NASA. (April 9, 2020). *International Space Station facts and figures.* https://www.nasa.gov/feature/facts-and-figures

[48] Collins Petersen, Carolyn. (July 19, 2019). *How long is a day on other planets?* ThoughtCo. https://www.thoughtco.com/day-length-other-planets-4165689

[49] The Planetary Society. (No date given). *Your guide to the International Space Station.* https://www.planetary.org/explore/space-topics/space-missions/iss.html

[50] Boyle, Rebecca. (March 9, 2012). How do you tell time on Mars? *Popular Science.* https://www.popsci.com/science/article/2012-03/how-do-you-tell-time-mars/

[51] Stromberg, Joseph. (August 9, 2012). How do you tell time on Mars? There's an app for that. *Smithsonian Magazine.* https://www.smithsonianmag.com/science-nature/how-do-you-tell-time-on-mars-theres-an-app-for-that-19350672/

[52] Tethers Unlimited, Inc. (No date given). *Connect the universe.* http://www.tethers.com/

[53]

Livingston, David. (No date given). *Dr. Robert Hoyt*. The Space Show. https:// www.thespaceshow.com/guest/dr.-robert-hoyt

[54] DARPA (Defense Advanced Research Projects Agency). (No date given). *Creating breakthrough technologies and capabilities for national security.* https://www.darpa.mil/

[55] Space Systems/Loral (SSL). (No date given). *Welcome to SSL.* http:// sslmda.com/html/about-us.php

[56] SBIR/STTR. (No date given). *The constructable platform: Modular architecture for a persistent platform in geo.* https://www.sbir.gov/sbirsearch/ detail/1250125   *{Phase I};*
SBIR/STTR. (No date given). *The constructable platform: Modular architecture for a persistent platform in geo.* https://www.sbir.gov/sbirsearch/ detail/1468003   *{Phase II}*

[57] Messier, Doug. (June 29, 2016). *DARPA awards TUI/Firmamentum contract to develop "constructable" persistent geo platform.* Parabolic Arc. http:// www.parabolicarc.com/2016/06/29/darpa-awards-tuifirmamentum-contract-develop-constructable-persistent-geo-platform/

[58] Hoyt, Rob. (December 13, 2017). *In-space recycling and manufacturing.* Tethers Unlimited/ Firmamentum. https://fiso.spiritastro.net/telecon/ Hoyt_12-13-17/Hoyt_12-13-17.pdf

[59] Tethers Unlimited Inc. (July 27, 2016). *Tethers unlimited wins DARPA grant for modular construction of geo platforms.* NewSpace Global. https://newspaceglobal.com/tethers-unlimited-wins-darpa-grant-modular-construction-geo-platforms/

[60] (Aug 31, 2013). *NASA-backed SpiderFab robot aims to build 3D-printed spaceship parts in orbit.* The Verge. https:// www.theverge.com/2013/8/31/4678046/nasa-funding-spiderfab-3d-printing-parts-in-space

[61] Hoyt, Robert, Jesse Cushing, Jeffrey Slostad. (July 8, 2013). *SpiderFab™ : Process for on-orbit construction of kilometerscale apertures.* Tethers Unlimited, Inc.   https://www.nasa.gov/sites/default/files/files/ Hoyt_2012_PhI_SpiderFab.pdf

[62] Chaitin, Daniel. (October 1, 2018). 7 fast facts on NASA's Lunar Gateway. *Washington Examiner.* https://www.washingtonexaminer.com/news/7-fast-facts-on-nasas-lunar-gateway

[63] NASA. (April 30, 2020). *Explore Moon to Mars.* https://www.nasa.gov/topics/moon-to-mars/overview

[64] NASA. (Last Updated: March 11, 2020). *Budget documents, strategic plans and performance reports.* http://www.nasa.gov/budget

[65] NASA. (February 13, 2018). *NASA's lunar outpost will extend human presence in deep space.* https://www.nasa.gov/feature/nasa-s-lunar-outpost-will-extend-human-presence-in-deep-space

[66] Foust, Jeff. (March 2, 2020). *First SLS launch now expected in second half of 2021.* SpaceNews. https://spacenews.com/first-sls-launch-now-expected-in-second-half-of-2021/

[67] SpaceX. (No date given). *Falcon Heavy.* https://www.spacex.com/vehicles/falcon-heavy/

[68] Purdue University College of Engineering. (Summer 2019). *NASA's lunar outpost will extend human presence in deep space.* Engineering Frontiers. https://engineering.purdue.edu/Frontiers/summer-2019/gateway

[69] NASA. (February 13, 2018). *NASA's Lunar Outpost will Extend Human Presence in Deep Space.* https://www.nasa.gov/feature/nasa-s-lunar-outpost-will-extend-human-presence-in-deep-space

[70] Sanders, Gerald and Michael Duke. (April 12, 2005). *In-situ resource utilization (ISRU) capability roadmap: Progress review.* Lunar and Planetary Institute. https://www.lpi.usra.edu/lunar_resources/documents/13_0IntegratedISRUPresenta.pdf

[71] ISRU Technology Development Project. (July 3, 2012). *Oxygen from regolith.* NASA. https://isru.nasa.gov/OxygenfromRegolith.html

[72] Barras, Colin. (October, 17 2008). Astronauts could mix DIY concrete for cheap moon base. *New Scientist.* https://www.newscientist.com/article/dn14977-astronauts-could-mix-diy-concrete-for-cheap-moon-base/

[73] Li, Gary, Danielle DeLatte, Jerome Gilleron, Samuel Wald, Therese Jones. (May 14, 2017). *Mining the Moon for rocket fuel to get us to Mars.* The Conversation. https://theconversation.com/mining-the-moon-for-rocket-fuel-to-get-us-to-mars-76123

[74] European Space Agency (ESA). (No date given). *Helium-3 mining on the lunar surface.* https://www.esa.int/Enabling_Support/

Preparing_for_the_Future/Space_for_Earth/Energy/
Helium-3_mining_on_the_lunar_surface;
NASA. (June 19, 2015). *Harnessing power from the Moon.* https://
www.nasa.gov/feature/harnessing-power-from-the-moon

[75] Mukunth, Vasudevan. (July 18, 2018). *Why are we going over the Moon on
ISRO and helium-3 all over again?* The Wire. https://thewire.in/the-sciences/
why-are-we-going-over-the-moon-on-helium-3-all-over-again

[76] Crawford, Ian A. (February 8, 2015). Lunar resources: A review. *Progress in Physical Geography: Earth and Environment, 39,* 137-167. https://
journals.sagepub.com/doi/pdf/10.1177/0309133314567585

[77] Rapp, Donald. (2013) The value of ISRU. In Donald Rapp
(Ed.), *Use of extraterrestrial resources for human space missions to
Moon or Mars.* (pp. 1-29). Springer-Verlag Berlin Heidelberg. https://
doi.org/10.1007/978-3-642-32762-9

[78] ISRUInfo. (No date given). *The space resources roundtable.* Retrieved May
24, 2020 from https://isruinfo.com/public/

[79] King, Hobart. (No date given). *REE - Rare earth elements and their uses.*
Geology.com. https://geology.com/articles/rare-earth-elements/

[80] Sohl, Frank, Breuer Doris. (2014) Differentiation, planetary. In Muriel Gargaud, William M. Irvine, Ricardo Amils, Henderson James
Cleaves, Daniele Pinti, José Cernicharo Quintanilla, Michel Viso (Eds.),
Encyclopedia of Astrobiology. Springer, Berlin, Heidelberg. https://
doi.org/10.1007/978-3-642-27833-4_430-2

[81] FutureTimeline.net. (September 1, 2018). *Launch costs to low Earth orbit,
1980-2100.* http://www.futuretimeline.net/data-trends/6.htm

[82] Byrne, Paul. (March 5, 2019). *Mining the Moon.* The Conversation. https://
theconversation.com/mining-the-moon-110744

[83] Fortin, Jacey and Karen Zraick. (March 25, 2019). First all-female spacewalk canceled because NASA doesn't have two suits that fit.
The New York Times. https://www.nytimes.com/2019/03/25/science/female-
spacewalk-canceled.html

[84] Bizony, Piers. (March 16, 2015). *The Problems of spacesuits 50 years after
the first spacewalk.* E&T. https://eandt.theiet.org/content/articles/2015/03/
the-problems-of-spacesuits-50-years-after-the-first-spacewalk/

[85] NASA. (April 7, 2002). *Spacewalking.* Human Space Flight. https://spaceflight.nasa.gov/shuttle/reference/faq/eva.html

[86] Patrick, Nancy, Joseph Kosmo, James Locke, Luis Trevino, Robert Trevino. (2011). Extravehicular activity operations and advancements. In Wayne Hale, Helen Lane, Gail Chapline, & Kamlesh Lulla (Eds.), *Wings In Orbit: Scientific and Engineering Legacies of the Space Shuttle 1971-2010* (pp. 110-129). Government Printing Office. https://www.nasa.gov/centers/johnson/pdf/584725main_Wings-ch3d-pgs110-129.pdf

[87] Howell, Elizabeth. (April 8, 2019). *Spacewalking astronauts battle stuck panel, wrangle cables on Space Station.* Space. https://www.space.com/spacewalking-astronauts-space-station-cable-upgrades-exp59.html

[88] Lucas, Jim. (February 7, 2018). *6 simple machines: Making work easier.* Live Science. https://www.livescience.com/49106-simple-machines.html

[89] Reisman, Garrett. (April 8, 2016). *What's it like to spacewalk at 17,500 mph?* (As told to A.J. Baime). The Drive. https://www.thedrive.com/travel/2928/whats-it-like-to-spacewalk-at-17-500-mph

[90] BBC. (August 6, 2008). *Can our TV signals be picked up on other planets?* BBC News Magazine. http://news.bbc.co.uk/2/hi/uk_news/magazine/7544915.stm

[91] Yiu, Yuen. (March 30, 2018). *How far can laser light travel?* Inside Science. https://www.insidescience.org/news/how-far-can-laser-light-travel

[92] Takenaka, Hideki , Alberto Carrasco-Casado, Mikio Fujiwara, Mitsuo Kitamura, Masahide Sasaki & Morio Toyoshima. (2017). Satellite-to-ground quantum-limited communication using a 50-kg-class microsatellite. *Nature Photon 11*, 502–508. https://doi.org/10.1038/nphoton.2017.107

[93] Book Series In Order. (No date given). *Expanse books in order.* https://www.bookseriesinorder.com/expanse/

[94] IMDB. (No date given). *The Expanse.* https://www.imdb.com/title/tt3230854/?ref_=tt_ov_inf

[95] NASA. (December 19, 2019). *Ceres.* NASA Science Solar System. https://solarsystem.nasa.gov/planets/dwarf-planets/ceres/in-depth/

[96] Universe Today. (No date given). *How do we colonize Ceres?* https://www.universetoday.com/143011/how-do-we-colonize-ceres/

[97] NASA. (July 30, 2018). *Top five technologies needed for a spacecraft to survive deep space.* https://www.nasa.gov/feature/top-five-technologies-needed-for-a-spacecraft-to-survive-deep-space

[98] Tahmaseb-McConatha, Jasmin. (Mar 14, 2015). Comforting third spaces. *Psychology Today.* https://www.psychologytoday.com/us/blog/live-long-and-prosper/201503/comforting-third-spaces

[99] https://www.relativityspace.com/

[100] Relativity [press release]. (May 6, 2019). *Relativity signs launch services agreement for multiple launches with spaceflight on Terran 1, world's first 3D printed rocket.* https://www.relativityspace.com/press-release/2019/06/29/relativity-signs-launch-services-agreement-for-multiple-launches-with-spaceflight-on-terran-1

[101] Scott, Clare. (December 12, 2018). *Relativity space 3D prints 11-foot-tall fuel tank with stargate 3D printer.* 3DPRINT.COM. https://3dprint.com/231703/relativity-space-3d-prints-fuel-tank/

[102] Relativity. (No date given). *Terran 1.* https://www.relativityspace.com/terran

[103] Ralph, Eric. (June 19, 2019). *SpaceX fires Falcon Heavy's 27 booster engines ahead of "most difficult launch ever".* Teslarati. https://www.teslarati.com/spacex-falcon-heavy-static-fire-most-difficult-launch-ever/

[104] Oberhaus, Daniel. (October 14, 2019). Massive, AI-powered robots are 3D-printing entire rockets. *Wired.* https://www.wired.com/story/massive-ai-powered-robots-are-3d-printing-entire-rockets/

[105] https://madeinspace.us/

[106] Knapp, Alex. (September 22, 2014). First zero-g 3D printer is on its way to the Space Station. *Forbes.* https://www.forbes.com/sites/alexknapp/2014/09/22/first-zero-g-3d-printer-is-on-its-way-to-the-space-station/

[107] Made In Space. (No date given). *Archinaut.* https://madeinspace.us/capabilities-and-technology/archinaut/

[108] Wang, Brian. (June 15, 2017). *Made in space also working on robotic manufacturing of large structures in space.* Next-

bigfuture. https://www.nextbigfuture.com/2017/06/made-in-space-also-working-on-robotic-manufacturing-of-large-structures-in-space.html

[109] Chrisman, Tim. (April 15, 2019). *A shield for whom?* Humanity in Space. https://future.humanityinspace.com/2019/04/15/private-security-or-private-army/

[110] Chrisman, Tim. (April 16, 2019). *The longest ride ever*. Humanity in Space. https://future.humanityinspace.com/2019/04/16/the-longest-ride-ever/

[111] FutureTimeline.net. (September 1, 2018). *Launch costs to low Earth orbit, 1980-2100*. http://www.futuretimeline.net/data-trends/6.htm

[112] Burke, Jack H. (June 13, 2019). China's new wealth-creation scheme: Mining the moon. *National Review*. https://www.nationalreview.com/2019/06/china-moon-mining-ambitious-space-plans/;
Malleta, King. (April 21, 2017). *India plans to start mining the moon by 2030*. NextShark. https://nextshark.com/india-plans-start-mining-moon-2030/;
Thornhill, James. (April 1, 2019). *Plan to mine the moon gives Australia opening in new space era*. Bloomberg. https://www.bloomberg.com/news/articles/2019-04-01/plan-to-mine-the-moon-gives-australia-opening-in-new-space-era

[113] Huebert, J. H. and Walter Block. (2007). Space environmentalism, property rights and the law. *University of Memphis Law Review, 37*, 281-309. http://www.walterblock.com/wp-content/uploads/publications/block-huebert_space-environmentalism-2006.pdf

[114] Apollo Lunar Landing Legacy Act. H.R.2617. 113th Congress (2013-2014). https://www.congress.gov/bill/113th-congress/house-bill/2617

[115] Sample, Ian. (July 19, 2019). Apollo 11 site should be granted heritage status, says space agency boss. *The Guardian*. https://www.theguardian.com/science/2019/jul/19/apollo-11-site-heritage-status-space-agency-moon

[116] Marshall Space Flight Center. (April 12, 2008). *Advanced space transportation program: Paving the highway to space*. NASA. https://www.nasa.gov/centers/marshall/news/background/facts/astp.html

[117] Chrisman, Tim. (April 29, 2019). *Could you print that for me?* Humanity in Space. https://future.humanityinspace.com/2019/04/29/could-you-

print-that-for-me/

[118] National Geographic. (November 3, 2015). Animals in space. https://www.nationalgeographic.com.au/space/animals-in-space.aspx

[119] Templeton, Graham. (June 27, 2014). *Geek answers: How do animals deal with zero gravity?* Geek. Archived Oct. 21, 2015. https://web.archive.org/web/20151021072910/https://www.geek.com/science/geek-answers-how-do-animals-deal-with-zero-gravity-1597647/

[120] CBC Radio. (April 26, 2019). *Mice reinvent the hamster wheel in zero gravity.* https://www.cbc.ca/radio/quirks/april-27-2019-oilsands-emissions-underestimated-chernobyl-s-wildlife-a-comet-trapped-in-an-asteroid-and-mo-1.5111304/mice-reinvent-the-hamster-wheel-in-zero-gravity-1.5111316

[121] NASAexplores. (April 10, 2009). *Animals in space.* NASA. https://www.nasa.gov/audience/forstudents/9-12/features/F_Animals_in_Space_9-12.html

[122] Seltzer, Adam. (No date given). *Zero-g mouse food dispenser.* NASA HUNCH. http://www.hunchdesign.com/uploads/2/2/0/9/22093000/z-g_mouse_food_dispenser.pdf;
NASA HUNCH. (No date given). *NASA HUNCH design and prototype projects for 2019-2020.* http://www.hunchdesign.com/uploads/2/2/0/9/22093000/projects_for_2019-2020.pdf

[123] Johnson, David Samuel. (December 22, 2016). The first fish in orbit. *Scientific American.* https://blogs.scientificamerican.com/guest-blog/the-first-fish-in-orbit/;
Reebs, Stéphan G. (2009). *Fishes in space.* howfishbehave.ca. http://www.howfishbehave.ca/pdf/Fishes%20in%20space.pdf

[124] Dickerson, Kelly. (October 24, 2013). *Animals born in space have a hard time adjusting to life on Earth.* Business Insider. https://www.businessinsider.com/space-born-animals-adjust-to-gravity-2013-10

[125] Monet, Dolores. (January 19, 2020). *History of clothing - Why we wear clothes.* Bellatory. https://bellatory.com/fashion-industry/History-of-Clothing-Why-We-Wear-Clothes

[126] Lady Gaga. (June 16, 2020). In *Wikipedia.* https://en.wikipedia.org/w/index.php?title=Lady_Gaga&oldid=962854879

[127] Jaeger, Max. (February 19, 2019). Legendary designer Karl Lagerfeld

dead at 85. *Page Six.* https://pagesix.com/2019/02/19/legendary-designer-karl-lagerfeld-dead-at-85/

[128] Miller, Meg. (August 8, 2016). *The extravagant, overlooked design of fashion show sets.* Fast Company. https://www.fastcompany.com/3062570/the-extravagant-overlooked-design-of-fashion-show-sets

[129] Fernholz, Tim. (March 17, 2015). *The space suit you could be wearing a few years from now is being designed in a small workshop in Brooklyn.* Quartz. https://qz.com/356649/the-space-suit-you-could-be-wearing-a-few-years-from-now-is-being-designed-in-a-small-workshop-in-brooklyn/; Harris, Mark. (September 2017). In pursuit of the perfect spacesuit. *Air & Space Magazine.* https://www.airspacemag.com/space/space-wear-180964337/

[130] NASA [@NASA]. (May 1, 2019). *What will astronauts wear on moonwalks in 5 years? We're currently working on options for our #Moon2024 mission + beyond.* [Tweet; link to image]. Twitter. https://twitter.com/nasa/status/1123667409343217666?lang=en

[131] Lyn Pesce, Nicole. (March 29, 2019). *NASA cancels all-female spacewalk because it doesn't have two spacesuits the right size.* MarketWatch. https://www.marketwatch.com/story/nasa-cancels-all-female-spacewalk-because-it-doesnt-have-two-spacesuits-the-right-size-2019-03-26

[132] NASA. (October 8, 2019). *A next generation spacesuit for the Artemis generation of astronauts.* https://www.nasa.gov/feature/a-next-generation-spacesuit-for-the-artemis-generation-of-astronauts

[133] Tousignant, Lauren. (February 10, 2017). Do we need to establish a police force in space? *New York Post.* https://nypost.com/2017/02/10/do-we-need-to-establish-a-police-force-in-space/

[134] Chrisman, Tim. (April 29, 2020). *More than just a gateway.* Humanity in Space. https://future.humanityinspace.com/2020/04/29/more-than-just-a-gateway/

[135] Chrisman, Tim. (April 23, 2019). *Semper paratus exteriores spatium.* Humanity in Space. https://future.humanityinspace.com/2019/04/23/semper-paratus-exteriores-spatium/

[136] O'Keefe, Caitlin. (August 28, 2015). *How we choose our near-earth asteroid targets.* Planetary Resources. https://www.planetaryresources.com/2015/08/how-we-choose-our-asteroid-targets/

[137] Glester, Andrew. (2018). The asteroid trillionaires. *Physics World, 31*(6), 33-35. https://physicsworld.com/a/the-asteroid-trillionaires/

[138] Rincon, Paul. (January 13, 2014). *Few asteroids are worth mining, suggests Harvard study*. BBC News. https://www.bbc.com/news/science-environment-25716103

[139] O'Keefe, Caitlin. (August 28, 2015). *How we choose our near-earth asteroid targets*. Planetary Resources. https://www.planetaryresources.com/2015/08/how-we-choose-our-asteroid-targets/

[140] Poponak, Noah, Matthew Porat, Michael Bishop, Chris Hallam, Heath P. Terry, Brett Feldman, Andrew Lee, Sam Wood, Peter Lapthorn, Gavin Parsons, Tais Correa, Adam Hotchkiss, David Tamberrino. (April 4, 2017). *Profiles in innovation-space: The next investment frontier*. The Goldman Sachs Group, Inc. https://www.gspublishing.com/content/research/en/reports/2017/04/04/49ead899-a05b-4822-9645-2948f2050a31.pdf

[141] UAE Space Agency. (No date given). *About UAE space agency*. https://www.space.gov.ae/Page/20120/20230/About-UAE-Space-Agency;
Taher, Ahmed. (January 11, 2019). Saudi Arabia and the space age: A dream becomes reality. *Majalla Magazine. 1730*, 8-13. https://issuu.com/majalla/docs/01__7_?e=1365300/67083595

[142] Treaty on principles governing the activities of states in the exploration and use of outer space, including the moon and other celestial bodies. United Nations Resolution Adopted By The General Assembly. 2222 (XXI). (1967). https://www.unoosa.org/oosa/en/ourwork/spacelaw/treaties/outerspacetreaty.html

[143] Australian Academy of Science. (December 17, 2018). *Mining the moon*. https://www.science.org.au/curious/space-time/mining-moon

[144] Foster, Craig. (2016). Excuse me, you're mining my asteroid: Space property rights and the U.S. space resource exploration and utilization act of 2015. *Journal of Law, Technology & Policy*, (2), 407-430. http://illinoisjltp.com/journal/wp-content/uploads/2016/11/Foster.pdf

[145] Crane, Leah. (October 7, 2017 [Online September 29, 2017]). Elon Musk's new plans for a moon base and a Mars mission by 2022. *New Scientist*. 3146 (1). https://www.newscientist.com/article/2149003-elon-musks-new-plans-for-a-moon-base-and-a-mars-mission-by-2022/;
Zubrin, Robert. (2018 Summer/Fall). Moon direct. *The New Atlantis*, (56),

14-47. https://www.thenewatlantis.com/publications/moon-direct

[146] Council on Foreign Relations. (November 30, 2017). *Who owns space? Commercial activity beyond Earth.* https://www.cfr.org/event/who-owns-space-commercial-activity-beyond-earth;
Paul, Deanna. (August 31, 2019). Space: The final legal frontier. *The Washington Post.* https://www.washingtonpost.com/technology/2019/08/31/space-final-legal-frontier/

[147] Australian Academy of Science. (December 17, 2018). *Mining the moon.* https://www.science.org.au/curious/space-time/mining-moon

[148] Cresswell, Matthew. (September 13, 2012). How Buzz Aldrin's communion on the moon was hushed up. *The Guardian.* https://www.theguardian.com/commentisfree/belief/2012/sep/13/buzz-aldrin-communion-moon#maincontent

[149] Teeter, Adam. (April 10, 2017). *Can you drink alcohol in space?* VinePair. https://vinepair.com/articles/can-you-drink-alcohol-in-space/

[150] Vinepair Staff. (No date given). *Sherry.* VinePair. https://vinepair.com/spirits-101/intro-sherry-guide/

[151] Bourland, Charles and Gregory L. Vogt. (2010). *The astronaut's cookbook: Tales, recipes, and more.* Springer.

[152] Kleinknecht, Kenneth. (August 10, 1972). *Memorandum to Chris Kraft, director of the Johnson Space Center.* NASA. p.9 of a faxed response to a 2006 Freedom of Information Act request by space historian Jennifer Ross-Nazzal. http://www.theblackvault.com/documents/space/skylabsherry.pdf

[153] Boyle, Alan. (October 14, 2010). *Alcohol in space? Da!* NBC News. https://www.nbcnews.com/science/alcohol-space-da-6C10403671

[154] Momken, Iman, Laurence Stevens, Audrey Bergouignan, Dominique Desplanches, Floriane Rudwill, Isabelle Chery, Alexandre Zahariev, Sandrine Zahn, T. Peter Stein, Jean Louis Sebedio, Estelle Pujos-Guillot, Maurice Falempin, Chantal Simon, Véronique Coxam, Tany Andrianjafiniony, Guillemette Gauquelin-Koch, Florence Picquet, and Stéphane Blanc. (October 2011). Resveratrol prevents the wasting disorders of mechanical unloading by acting as a physical exercise mimetic in the rat. *The FASEB Journal 25*(10), 3646-3660. https://www.fasebj.org/doi/full/10.1096/fj.10-177295

[155] Teeter, Adam. (April 10, 2017). *Can you drink alcohol in space?* VinePair. https://vinepair.com/articles/can-you-drink-alcohol-in-space/

[156] NASA. (September 21, 2001). *Suds in space.* https://science.nasa.gov/science-news/science-at-nasa/2001/ast21sep_1/

[157] Lumsden, Bill. (2015). *The impact of micro-gravity on the release of oak extractives into spirit.* Ardbeg Distillery. https://www.ardbeg.com/sites/ardbeg.com/files/2017-10/ARD9109SupernovaWhitePaperA4.pdf;

[158] Davison, Anna. (July 31, 2007). Beer in space: A short but frothy history. *New Scientist.* https://www.newscientist.com/article/dn12388-beer-in-space-a-short-but-frothy-history/

[159] Vostok Space Beer. (No date given). *Vostok space beer: The world's first beer for space.* https://vostokspacebeer.com/

[160] Hsu, Jeremy. (May 18, 2011). *Beer for space tourists: More taste, fewer wet burps.* Space. https://www.space.com/11706-space-beer-microgravity.html

[161] Hongo, Jun. (July 31, 2015). Suntory plans space-aged whisky. *Wall Street Journal.* https://blogs.wsj.com/japanrealtime/2015/07/31/suntory-plans-space-aged-whisky/

[162] Collins, William, Henry Mertens, and E. Amold Higgins. (1985). *Some effects of alcohol and simulated altitude on complex performance scores and breathalyzer readings.* Federal Aviation Administration. https://www.faa.gov/data_research/research/med_humanfacs/oamtechreports/1980s/media/AM85-05.pdf

[163] Medical Dictionary. (No date given). *Think-drink effect.* The Free Dictionary by Farlex. https://medical-dictionary.thefreedictionary.com/think-drink%20effect

[164] Lufkin, Bryan. (February 20, 2017). *Why astronauts are banned from getting drunk in space.* BBC Future. https://www.bbc.com/future/article/20170217-why-astronauts-are-banned-from-getting-drunk-in-space

[165] Chapman, Gray. (October 10, 2017). *What's a spinzall, and does your bar need one?* SevenFifty Daily. https://daily.sevenfifty.com/whats-a-spinzall-and-does-your-bar-need-one/

[166] Chrisman, Tim. (May 24, 2019). *Dude hold my beer.* Humanity in Space. https://future.humanityinspace.com/2019/05/24/dude-hold-my-beer/

[167] Archibald, Anna. (June 9, 2020). Meet the mad scientist of bartending. https://www.winemag.com/2019/04/01/dave-arnold-mad-scientist-bartender/

[168] https://www.amazon.com/Centrifuge-Four-Es-Scientific-Benchtop/

dp/B0859XCLTG

[169] Woodford, Chris. (December 15, 2019). *Centrifuges*. Explain That Stuff! https://www.explainthatstuff.com/centrifuges.html.

[170] Wolf, Buck. (February 23, 2006). *'Misunderestimated' Bushisms 'resignate'*. ABC News. https://abcnews.go.com/Entertainment/WolfFiles/story?id=90907&page=1

[171] United States Geological Survey (USGS). (No date given). *Water in space: How does water behave in outer space?* https://www.usgs.gov/special-topic/water-science-school/science/water-space-how-does-water-behave-outer-space?qt-science_center_objects=0#qt-science_center_objects

[172] Royal Museums Greenwich. (No date given). *What do astronauts eat in space?* https://www.rmg.co.uk/discover/explore/space-stargazing/space-exploration/what-do-astronauts-eat-in-space

[173] Cosmic Lifestyle Corporation. (April 2, 2015). *Zero gravity cocktail project*. Kickstarter. https://www.kickstarter.com/projects/spacemansam/zero-gravity-cocktail-project

[174] Ballantine's. (September 2, 2015). *Introducing the Ballantine's space glass*. Medium. https://medium.com/space-glass/introducing-the-ballantine-s-space-glass-d7722772f9a7#.e63pu6kd4;
Maison Mumm. (No date given). *Mumm Grand Cordon Stellar*. https://www.mumm.com/en/mumm-grand-cordon-stellar;
Stratasys Direct Manufacturing. (No date given). *Case study: 3D printing brings the familiarity of home to the international space station.*

[175] Gordon, Ilana. (January 11, 2017). *The science of having sex in space*. Medium. https://medium.com/omgfacts/the-science-of-having-sex-in-space-220874f792bb;
Koerth, Maggie. (March 14, 2017). *Space sex is serious business*. FiveThirtyEight. https://fivethirtyeight.com/features/space-sex-is-serious-business/

[176] Hodge, Mark. (December 7, 2018). Sex in space would be a nightmare, scientist says. *New York Post*. https://nypost.com/2018/12/07/sex-in-space-would-be-a-nightmare-scientist-says/

[177] Aldrich, Clayton. (No date given). *These crazy people just became the first couple to have sex while skydiving*. Boredom Therapy. https://boredomtherapy.com/skydiving-sex/

[178] Meggs, Lori. (September 5, 2007). *Cardiovascular system gets 'lazy' in space; New study gets blood flowing on station*. NASA. https://www.nasa.gov/

mission_pages/station/research/cciss_feature.html

[179] Patel, Neel V. (December 21, 2016). *What happens to dicks in space?* Inverse. https://www.inverse.com/article/25523-penis-dick-in-space-nasa-gravity

[180] Noonan, Raymond J. (2004). Outer Space and Antarctica: Sexuality factors in extreme environments. In Robert T. Francoeur & Raymond J. Noonan (Eds.), *The continuum complete international encyclopedia of sexuality* (pp. 795-812). Continuum. https://kinseyinstitute.org/pdf/ccies-outerspaceantarctica.pdf

[181] European Space Agency. (No date given). *Living in space.* https://www.esa.int/Science_Exploration/Human_and_Robotic_Exploration/Astronauts/Living_in_space

[182] Boyle, Alan. (July 24, 2006). *Outer-space sex carries complications.* NBC News. http://www.nbcnews.com/id/14002908/ns/technology_and_science-space/t/outer-space-sex-carries-complications/; Freitas, Robert A., Jr. (April 1983). Sex in space. *Sexology Today*, 48, 58-64. http://www.rfreitas.com/Astro/SexxxInSpace.htm

[183] Koziol, Michael. (November 3, 2016). 10 disgusting ways your body betrays you in space: Sudden peeing, sweat balls, and so much more. *Popular Science.* https://www.popsci.com/how-your-body-betrays-you-in-space/

[184] Roach, Mary. (2010). *Packing for Mars : The curious science of life in the void.* W. W. Norton.

[185] Scaturro, Giorgia. (April 30, 2009). A two-seater suit for space-lovers. *Wired.* https://www.wired.co.uk/article/a-two-seater-suit-for-space-lovers

[186] Bryner, Jeanna. (July 7, 2008). *For better or worse, sex in space is inevitable.* Space. https://www.space.com/5594-worse-sex-space-inevitable.html

[187] Gaming Club Casino Blog. (February 11, 2018). *Top games for a trip to Mars.* https://blog.gamingclub.com/top-games-trip-mars/

[188] Sortal, Nick. (August 22, 2017). Dealer-less poker tables? Electronic alternative is faster — and you don't need to tip. *Miami Herald.* https://www.miamiherald.com/entertainment/article168691777.html

[189] Olsen, Erik. (November 21, 2014). *SERE training develops leaders for complex environment.* U.S. Army. https://www.army.mil/article/138765/sere_training_develops_leaders_for_complex_environment

[190] Sleep Advisor. (April 3, 2020). *45 sleep mantras better than counting sheep.* https://www.sleepadvisor.org/sleep-mantras/

[191] Plumer, Brad. (March 1, 2016). *Sleeping in space is weirdly difficult — And other lessons from Scott Kelly's year in orbit.* Vox. https://www.vox.com/2016/1/30/10873616/nasa-scott-kelly-space-station

[192] Basner, Mathias and David F Dinges. (September 1, 2014). Lost in space: Sleep. *The Lancet, 13* (9), 860-862. https://doi.org/10.1016/S1474-4422(14)70176-0

[193] NASA. (Jan 7, 2016). *NASA Research Reveals Biological Clock Misalignment Effects on Sleep for Astronauts.* https://www.nasa.gov/feature/ames/nasa-research-reveals-biological-clock-misalignment-effects-on-sleep-for-astronauts

[194] NASA. (January 7, 2016). *NASA research reveals biological clock misalignment effects on sleep for astronauts.* https://www.nasa.gov/feature/ames/nasa-research-reveals-biological-clock-misalignment-effects-on-sleep-for-astronauts

[195] Hollingham, Richard. (May 10, 2017). *The quest to help astronauts sleep better.* BBC Future. https://www.bbc.com/future/article/20170509-the-quest-to-help-astronauts-sleep-better

[196] Basner, Mathias, David F. Dinges, Daniel Mollicone, Adrian Ecker, Christopher W. Jones, Eric C. Hyder, Adrian Di Antonio, Igor Savelev, Kevin Kan, Namni Goel, Boris V. Morukov, and Jeffrey P. Sutton. (February 12, 2013). Mars 520-d mission simulation reveals protracted crew hypokinesis and alterations of sleep duration and timing. *PNAS, 110*(7), 2635-2640. https://doi.org/10.1073/pnas.1212646110

[197] Czeisler, Charles A., Jeanne F. Duffy, Theresa L. Shanahan, Emery N. Brown, Jude F. Mitchell, David W. Rimmer, Joseph M. Ronda, Edward J. Silva, James S. Allan, Jonathan S. Emens, Derk-Jan Dijk, Richard E. Kronauer. (June 25, 1999). Stability, precision, and near-24-hour period of the human circadian pacemaker. *Science, 284*(5423), 2177-2181. https://doi.org/10.1126/science.284.5423.2177

[198] Weitering, Hanneke. (March 23, 2017). *Sleepless in space: Therapy helps astronauts snooze.* Space. https://www.space.com/36165-therapy-helps-astronauts-sleep.html;
Wu, Bin, Yue Wang, Xiaorui Wu, Dong Liu, Dong Xu, and Fei Wang. (2018). On-orbit sleep problems of astronauts and

countermeasures. *Military Medical Research, 5*(1), 1-12. https://dx.doi.org/10.1186%2Fs40779-018-0165-6

[199] Howard, Jenny. (December 15, 2016). *Seven ways astronauts improve sleep may help you snooze better on Earth.* NASA. https://www.nasa.gov/mission_pages/station/research/astronauts_improve_sleep

[200] Zhou, Eric, Paula Gardiner, Suzanne M. Bertisch. (June 2017). Integrative medicine for insomnia. *Medical Clinics of North America, 101*(5), 865–879. https://doi.org/10.1016/j.mcna.2017.04.005

[201] Garrett-Bakelman, Francine E., Manjula Darshi, Stefan J. Green, Ruben C. Gur, Ling Lin, Brandon R. Macias, Miles J. McKenna, Cem Meydan, Tejaswini Mishra, Jad Nasrini, Brian D. Piening, Lindsay F. Rizzardi, Kumar Sharma, Jamila H. Siamwala, Lynn Taylor, Martha Hotz Vitaterna, Maryam Afkarian, Ebrahim Afshinnekoo, Sara Ahadi6 Aditya Ambati, Maneesh Arya, Daniela Bezdan, Colin M. Callahan, Songjie Chen, Augustine M. K. Choi, George E. Chlipala, Kévin Contrepois, Marisa Covington, Brian E. Crucian. Immaculata De Vivo, David F. Dinges, Douglas J. Ebert, Jason I. Feinberg, Jorge A. Gandara, Kerry A. George, John Goutsias, George S. Grills, Alan R. Hargens, Martina Heer, Ryan P. Hillary, Andrew N. Hoofnagle, Vivian Y. H. Hook, Garrett Jenkinson, Peng Jiang, Ali Keshavarzian, Steven S. Laurie, Brittany Lee-McMullen, Sarah B. Lumpkins, Matthew MacKay, Mark G. Maienschein-Clin, Ari M. Melnick, Tyler M. Moore, Kiichi Nakahira, Hemal H. Patel, Robert Pietrzyk, Varsha Rao, Rintaro Saito, Denis N. Salins, Jan M. Schilling, Dorothy D. Sears, Caroline K. Sheridan, Michael B. Stenger, Rakel Tryggvadottir, Alexander E. Urban, Tomas Vaisar, Benjamin Van Espen, Jing Zhang, Michael G. Ziegler, Sara R. Zwart, John B. Charles, Craig E. Kundrot, Graham B. I. Scott, Susan M. Bailey, Mathias Basner, Andrew P. Feinberg, Stuart M. C. Lee, Christopher E. Mason, Emmanuel Mignot, Brinda K. Rana, Scott M. Smith, Michael P. Snyder, Fred W. Turek. (April 12, 2019). The NASA twins study: A multidimensional analysis of a year-long human spaceflight. *Science, 364*(6436), 643-63. https://science.sciencemag.org/content/364/6436/eaau8650

[202] Kaplan, Sarah. (March 16, 2018). The truth about Astronaut Scott Kelly's viral 'space genes'. *The Washington Post.* https://www.washingtonpost.com/news/speaking-of-science/wp/2018/03/16/the-truth-about-astronaut-scott-kellys-viral-space-genes/

[203] Parihar, Vipan K., Barrett Allen, Katherine K. Tran, Trisha G. Macaraeg, Esther M. Chu, Stephanie F. Kwok, Nicole N. Chmielewski, Brianna M. Craver, Janet E. Baulch, Munjal M. Acharya, Francis A. Cucinotta and Charles L.

Limoli. (May 1, 2015). What happens to your brain on the way to Mars. *Science Advances*, 1(4), e1400256. https://doi.org/10.1126/sciadv.1400256

[204] Achenbach, Joel. (April 11, 2019). NASA Kelly twins study shows harsh effects of space flight and a brutal return to earth. *The Washington Post*. https://www.washingtonpost.com/science/2019/04/11/kelly-twin-astronauts-study-shows-harsh-effects-space-flight-brutal-return-earth/

[205] Chrisman, Tim. (April 18, 2019). *Habitat fever*. Humanity in Space. https://future.humanityinspace.com/2019/04/18/habitat-fever/

[206] Seitz, Dan. (January 18, 2019). Yes, cabin fever is real—here's how to prevent it. *Popular Science*. https://www.popsci.com/prevent-cabin-fever/

[207] Tarvainen, Sinikka. (November 14, 2019). *Unprecedented protests reveal 'fragile' social order, expert says.* dpa International. https://www.dpa-international.com/topic/unprecedented-protests-reveal-fragile-social-order-expert-urn%3Anewsml%3Adpa.com%3A20090101%3A191115-99-737674

[208] Carr, Craig L. (October 1990). Tacit consent. *Public Affairs Quarterly*, 4(4), 335-345.

[209] Barkan, Steven E. (2012). Chapter 5: Social structure and social interaction. In Steven E. Barkan, *Sociology: Comprehensive Edition* (v. 1.0). unnamed publisher (see https://2012books.lardbucket.org/ for details). https://2012books.lardbucket.org/books/sociology-comprehensive-edition/s08-social-structure-and-social-in.html

[210] Astrosociology Research Institute (ARI). (No date given). *About ARI*. http://www.astrosociology.org/aboutARI.html

[211] Astrosociology Research Institute (ARI). (No date given). *The Journal of Astrosociology (JOA)*. http://www.astrosociology.org/joa.html

[212] Brown, Mike. (April 14, 2019). *Mars Colony: A city on Mars could descend into cabin fever and nationalism*. Inverse. https://www.inverse.com/article/54813-mars-colony

[213] Pass, Jim. (2004). *Inaugural essay: The definition and relevance of astrosociology in the twenty-first century (Part 2: Relevance of astrosociology as a new subfield of sociology)*. Astrosociology Research Institute. http://astrosociology.com/Library/Iessay/iessay_p2.pdf

[214] NASA. (July 23, 2018). *Top five teams win a share of $100,000 in virtual modeling stage of NASA's 3D-printed habitat competition*. https://www.nasa.gov/directorates/spacetech/centennial_challenges/3DPHab/five-teams-win-a-share-of-100000-in-virtual-modeling-stage

[215] Solomon, Scott. (No date given). *About*. Rice University. https://solomon.rice.edu/

[216] Solomon, Scott. (January 17, 2018). *Evolutionary biology on Mars* [Video]. TED. https://www.youtube.com/watch?v=Z2jlWzoVPFc

[217] https://www.doctorwho.tv/

[218] Chrisman, Tim. (April 29, 2020). *More than just a Gateway*. Humanity in Space. https://future.humanityinspace.com/2020/04/29/more-than-just-a-gateway/

[219] Chrisman, Tim. (April 26, 2020). *Buzz's cycler*. Humanity in Space. https://future.humanityinspace.com/2020/04/26/buzzs-cycler/

[220] Wakabayashi, Daisuke. (March 19, 2018). Self-driving uber car kills pedestrian in Arizona, where robots roam. *The New York Times*. https://www.nytimes.com/2018/03/19/technology/uber-driverless-fatality.html

[221] Brody, Jane E. (May 11, 2020). A pandemic benefit: The expansion of telemedicine. *The New York Times*. https://www.nytimes.com/2020/05/11/well/live/coronavirus-telemedicine-telehealth.html

[222] Wicklund, Eric. (October 25, 2017). *Telemedicine robots: Out of science fiction and into the mainstream*. mHealthIntelligence. https://mhealthintelligence.com/features/can-telemedicine-robots-move-from-fantasy-to-fact

[223] Choi, Paul, Rod Oskouian, and R. Shane Tubbs. (May 2018). Telesurgery: Past, present, and future. *Cureus*, *10*(5): e2716. https://www.ncbi.nlm.nih.gov/pmc/articles/PMC6067812/

[224] [skyluke89]. (Jul 20, 2014). *How far does light travel in 1 millisecond that is, in one hundredth of a scecond?* [Online forum post]. Brainly. https://brainly.com/question/85844

[225] Howell, Elizabeth. (April 24, 2015). *What Is a Geosynchronous Orbit?* Space. https://www.space.com/29222-geosynchronous-orbit.html

[226] Sneath, Evan, Christopher Korte and Grant Schaffner. (January 5, 2020). *Semi-autonomous robotic surgery for space exploration missions*. AIAA Scitech 2020 Forum. https://doi.org/10.2514/6.2020-1379;

Yip, Michael and Nikhil Das. (2018). Robot autonomy for surgery. Chapter 10 in Rajni Patel (Ed.), *The Encyclopedia of Medical Robotics, Volume 1: Minimally Invasive Surgical Robotics* (pp. 281-313). World Scientific. https://doi.org/10.1142/9789813232266_0010

[227] Howell, Elizabeth. (June 5, 2017). *Here's what emergency medicine will look like for astronauts in space.* Space. https://www.space.com/37097-heres-what-emergency-medicine-will-look-like-for-astronauts-in-space.html

[228] Goldhahn, Jörg, Vanessa Rampton, Giatgen A. Spinas. (November 7, 2018). Could artificial intelligence make doctors obsolete? *BMJ*, 363, k4563. https://doi.org/10.1136/bmj.k4563;
Olson, Parmy. (June 28, 2018). This AI just beat human doctors on a clinical exam. *Forbes.* https://www.forbes.com/sites/parmyolson/2018/06/28/ai-doctors-exam-babylon-health/;
Svoboda, Elizabeth. (September 25, 2019). Your robot surgeon will see you now. *Nature* 573, S110-S111. https://www.nature.com/articles/d41586-019-02874-0

[229] Stasis (fiction). (15 June 2020). In *Wikipedia.* https://en.wikipedia.org/w/index.php?title=Stasis_(fiction)&oldid=962728513

[230] Johns Hopkins Medicine. (No date given). *Therapeutic hypothermia after cardiac arrest.* https://www.hopkinsmedicine.org/health/treatment-tests-and-therapies/therapeutic-hypothermia-after-cardiac-arrest

[231] Villazon, Luis. (No date given). *How long can you be kept on a life support machine?* Science Focus. https://www.sciencefocus.com/the-human-body/how-long-can-you-be-kept-on-a-life-support-machine/

[232] Philip Jaekl. (December 14, 2017). *In cold blood.* Aeon. https://aeon.co/essays/how-freezing-patients-could-save-lives-and-even-reverse-death

[233] Cross, Mary. (June 27, 2017). *Quick cooling thermosuit may prevent brain damage in stroke patients.* Tulane News. https://news.tulane.edu/news/quick-cooling-thermosuit-may-prevent-brain-damage-stroke-patients

[234] Rosen, Rebecca J. (October 3, 2012). Blood in zero gravity: NASA tries to prepare for surgery in space. *The Atlantic.* https://www.theatlantic.com/technology/archive/2012/10/blood-in-zero-gravity-nasa-tries-to-prepare-for-surgery-in-space/263171/

[235] Rogers, Adam. (July 24, 2017). Zero-g blood and the many horrors of space surgery. *Wired.* https://www.wired.com/story/zero-g-blood-and-the-many-horrors-of-space-surgery/

[236] Cristoforetti, Samantha. (April 29, 2012). *No molecule shall stand still!* European Space Agency (ESA). https://blogs.esa.int/astronauts/2012/04/29/no-molecule-shall-stand-still/

[237] Kruszelnicki, Karl S. [Dr Karl]. (June 9, 2015). *Dr Karl's great moments in science: How do astronauts breathe in space?* ABC Science. https://www.abc.net.au/science/articles/2015/06/09/4249936.htm

[238] NASA. (June 28, 2017). *Space station regenerative ECLSS flow diagram.* https://www.nasa.gov/offices/oct/image-feature/space-station-regenerative-eclss-flow-diagram

[239] NASA. (November 1, 2000). *Water on the Space Station.* https://science.nasa.gov/science-news/science-at-nasa/2000/ast02nov_1

[240] NASA. (November 12, 2000). *Breathing easy on the Space Station.* https://science.nasa.gov/science-news/science-at-nasa/2000/ast13nov_1

[241] Radford, Benjamin. (June 6, 2016). *Is it safe to drink blood?* Live Science. https://www.livescience.com/15899-drinking-blood-safe.html

[242] Ferreira, Becky. (Aug 9, 2018). *Mike Pence says America's Space Force will seek peace through strength.* Vice. https://www.vice.com/en_us/article/7xqdez/mike-pence-says-americas-space-force-will-seek-peace-through-strength

[243] McKinley, Cynthia A. S. (Spring 2000). The guardians of space: Organizing America's space assets for the twenty-first century. *Aerospace Power Journal, XIV*(1), 37-45. https://www.spacefaringamerica.com/wp-content/uploads/2018/06/McKinley-Guardians-of-Space-ADA519071.pdf

[244] Bennett, James C. (Winter 2011). Proposing a 'Coast Guard' for space. *The New Atlantis, 30,* 50-68. https://www.thenewatlantis.com/publications/proposing-a-coast-guard-for-space

[245] Treaty on principles governing the activities of states in the exploration and use of outer space, including the moon and other celestial bodies. United Nations Resolution Adopted By The General Assembly. 2222 (XXI). (1967). https://www.unoosa.org/oosa/en/ourwork/spacelaw/treaties/outerspacetreaty.html

[246] Sinclair, Michael. (May 21, 2018). *The US needs a 'Coast Guard' for space: Semper paratus exteriores spatium.* Breaking Defense. https://breakingdefense.com/2018/05/the-us-needs-a-coast-guard-for-space-semper-paratus-exteriores-spatium/

[247] https://www.uscg.mil/

[248] U.S. Coast Guard Historian's Office. (No date given). *Missions*. U.S. Coast Guard. https://www.history.uscg.mil/Home/Missions/

[249] Werner, Debra. (October 4, 2019). *EU space envoy calls for satellites to leave orbit soon after mission ends*. SpaceNews. https://spacenews.com/eu-space-envoy-calls-for-satellites-to-leave-orbit-soon-after-mission-ends/

[250] NASA. (October 13, 2014). *Water recycling*. https://www.nasa.gov/content/water-recycling/

[251] European Space Agency (ESA). (No date given). PR, media and communications. https://www.esa.int/About_Us/EAC/PR_Media_and_Communications

[252] European Space Agency (ESA). (Sep 01, 2017). *Recycling water on the ISS: How it works pt. 1*. https://www.esa.int/ESA_Multimedia/Images/2017/09/Recycling_water_on_the_ISS_How_it_works_pt._1

[253] European Space Agency (ESA). (Sep 01, 2017). *Recycling water on the ISS: How it works pt. 2*. https://www.esa.int/ESA_Multimedia/Images/2017/09/Recycling_water_on_the_ISS_How_it_works_pt._2

[254] Chrisman, Tim. (April 26, 2020). *Buzz's cycler*. Humanity in Space. https://future.humanityinspace.com/2020/04/26/buzzs-cycler/

[255] Encyclopedia.com. (July 9 2020). Banishment. In *Encyclopedia.com*. https://www.encyclopedia.com/social-sciences-and-law/law/law/banishment

[256] Fischer , David Hackett. (1989). *Albion's seed: Four British folkways in America*. Oxford University Press.;
New England Historical Society. (No date given). *The great migration of picky Puritans, 1620-40*. https://www.newenglandhistoricalsociety.com/great-migration-of-picky-puritans-1620-40/

[257] CBS. (No date given). *Survivor*. https://www.cbs.com/shows/survivor/

[258] Discovery. (No date given). *Naked and Afraid*. https://www.discovery.com/shows/naked-and-afraid

[259] U.S. Mission China. (2018). *China 2018 human rights report*. https://china.usembassy-china.org.cn/china-2018-human-rights-report/

[260] Podvig, Pavel, and Hui Zhang. (January 2008). *Russian and Chinese responses to U.S. military plans in space*. American Academy of Arts & Sciences. https://www.amacad.org/publication/russian-and-chinese-

responses-us-military-plans-space;
Tucker, Patrick. (March 14, 2019). Pentagon wants to test a space-based weapon in 2023. Defense One. https://www.defenseone.com/technology/2019/03/pentagon-wants-test-space-based-weapon-2023/155581/

[261] Lawrence Livermore National Laboratory. (No date given). *The research library*. https://library.llnl.gov/

[262] Wood, Daniel. (March 6, 2014). *Space-based solar power*. Department of Energy. https://www.energy.gov/articles/space-based-solar-power

[263] Gdoutos, Eleftherios, Christophe Leclerc, Fabien Royer, Michael D. Kelzenberg, Emily C. Warmann, Pilar Espinet-Gonzalez, Nina Vaidya, Florian Bohn, Behrooz Abiri, Mohammed R. Hashemi, Matan Gal-Katziri, Austin Fikes, Harry Atwater, Ali Hajimiri and Sergio Pellegrino. (2018). *A lightweight tile structure integrating photovoltaic conversion and rf power transfer for space solar power applications*. Paper AIAA 2018-2202 presented at the American Institute of Aeronautics & Astronautics Conference and Exposition. https://doi.org/10.2514/6.2018-2202

[264] Caltech. (No date given). *Caltech space solar power project*. https://www.spacesolar.caltech.edu/how-it-works/

[265] Conners, Deanna. (July 29, 2018). *Space-based solar power: How close to reality?* EarthSky. https://earthsky.org/earth/space-based-solar-energy-power-getting-closer-to-reality

[266] Division of Engineering and Applied Science, California Institute of Technology. (Fall 2015). The space solar power initiative. *ENGenious, 12*, 8-12. https://caltechcampuspubs.library.caltech.edu/2931/1/2015-ENGenious12.pdf

[267] Wood, Daniel. (March 6, 2014). *Space-based solar power*. Department of Energy. https://www.energy.gov/articles/space-based-solar-power

[268] Lawrence Livermore National Laboratory. (No date given). *The research library*. https://library.llnl.gov/

[269] Pombriant, Denis. (June 21, 2019). *SBSP–Solar power from space*. Medium. https://medium.com/@DenisPombriant/sbsp-solar-power-from-space-f612e7bc39d8

[270] Baiocchi, Dave and William Welser IV. (May/June 2015). The democratization of space: New actors need new rules. *Foreign Affairs 94* (3), 98-104.

https://www.foreignaffairs.com/articles/space/2015-04-20/
democratization-space

[271] Miller, Gregory. (Fall 2019). Space pirates, geosynchronous guerrillas, and nonterrestrial terrorists: Nonstate threats in space. *Air and Space Power Journal, 33*(3), 33-51. https://www.airuniversity.af.edu/Portals/10/ASPJ/journals/Volume-33_Issue-3/F-Miller.pdf

[272] McFate, Sean. (2019). *Mercenaries and war: Understanding private armies today.* National Defense University Press. https://ndupress.ndu.edu/Portals/68/Documents/strat-monograph/mercenaries-and-war.pdf

[273] Gaultier, Leonard, Garine Hovsepian, Ayesha Ramachandran, Ian Wadley and Badr Zerhdoud. (January 2001). *The mercenary issue at the UN commission on human rights: The need for a new approach.* International Alert. http://www.operationspaix.net/DATA/
DOCUMENT/6041~v~The_Mercenary_Issue_at_the_UN_Commission_on-_Human_Rights.pdf

[274] Glabau, Danya. (February 15, 2019). *Cyborgs at the frontiers.* (originally presented as a talk at the Transpecies Society in Barcelona, Spain on January 13, 2019, titled, "Cyborgs at the frontiers: Cybernetic speculation and human transendence"). https://danyaglabau.com/2019/02/15/cyborgs-at-the-frontiers/